Agriculture and Employment in Developing Countries

Westview Special Studies

The concept of Westview Special Studies is a response to the continuing crisis in academic and informational publishing. Library budgets for books have been severely curtailed. Ever larger portions of general library budgets are being diverted from the purchase of books and used for data banks, computers, micromedia, and other methods of information retrieval. Interlibrary loan structures further reduce the edition sizes required to satisfy the needs of the scholarly community. Economic pressures on university presses and the few private scholarly publishing companies have greatly limited the capacity of the industry to properly serve the academic and research communities. As a result, many manuscripts dealing with important subjects, often representing the highest level of scholarship, are no longer economically viable publishing projects--or, if accepted for publication, are typically subject to lead times ranging from one to three years.

Westview Special Studies are our practical solution to the problem. As always, the selection criteria include the importance of the subject, the work's contribution to scholarship, and its insight, originality of thought, and excellence of exposition. We accept manuscripts in camera-ready form, typed, set, or word processed according to specifications laid out in our comprehensive manual, which contains straightforward instructions and sample pages. The responsibility for editing and proofreading lies with the author or sponsoring institution, but our editorial staff is always available to answer questions and provide guidance.

The result is a book printed on acid-free paper and bound in sturdy, library-quality soft covers. We manufacture these books ourselves using equipment that does not require a lengthy make-ready process and that allows us to publish first editions of 300 to 1000 copies and to reprint even smaller quantities as needed. Thus, we can produce Special Studies quickly and can keep even very specialized books in print as long as there is a demand for them.

About the Book and Author

High rates of growth in agricultural production need not be incompatible with increased employment, income, and the satisfaction of basic needs in the lower-income developing countries. Emphasizing this theme, the author presents three alternative agricultural development strategies and suggests guidelines for identifying appropriate policies and programs. These policies are designed to maximize agricultural production and provide additional opportunities for employment, thereby increasing the purchasing power of rural families.

The three models of agricultural development (the Dual-Size Structure model, the Uniformly Small Farm model, and the Mixed Characteristics model) differ in terms of ownership and access to land and other resources, macroeconomic policies affecting relative prices, and sectoral policies that determine the type of technologies available to farmers. The author assesses the potential of each strategy for generating employment and increasing income, drawing on case studies of Latin American, Asian, and African economies.

Dr. Bela Mukhoti is an agricultural economist in the International Economics Division, Economic Research Service of the U.S. Department of Agriculture.

Agriculture and Employment in Developing Countries
Strategies for Effective Rural Development

Bela Mukhoti

Westview Press / Boulder and London

Westview Special Studies in Social, Political, and Economic Development

The views presented in this book are those of the author and do not necessarily represent those of the institution with which she is associated.

Published in 1985 in the United States of America by Westview Press, Inc., 5500 Central Avenue, Boulder, Colorado 80301; Frederick A. Praeger, Publisher

Library of Congress Cataloging in Publication Data
Mukhoti, Bela.
    Agriculture and employment in developing countries.
    (Westview special studies in social, political, and economic development.)
    Includes bibliographical references.
    1. Agricultural laborers--Developing countries--Supply
and demand.   2. Agriculture and state--Developing
countries.   3. Rural development--Government policy--
Developing countries.   4. Food supply--Government policy
--Developing countries.   5. Developing countries--Full
employment policies.   I. Title.
HD1542.M85 1985   331.12'53'091724   84-25710
ISBN 0-8133-7032-9

Printed and bound in the United States of America

10 9 8 7 6 5 4 3 2 1

Contents

Acknowledgements

This book is based on a project report written in 1982 under a participating agency agreement between the Agency for International Development and the U.S. Department of Agriculture under my project leadership. Many of the ideas in the AID report drew heavily on my previous work. Included in my previous work are my doctoral dissertation, journal articles published in the 1960s and 1970s dealing with agrarian structure and its relationship to agricultural development, and, in particular, an unpublished paper on "Patterns of Technological Transformation of Agriculture and Economic Development" prepared for presentation at the 1980 Allied Social Science Convention in Denver, Colorado.

I wish to thank Dr. Bruce Johnston for agreeing to collaborate on the AID project. He used his extensive knowledge of the development arena and made an invaluable contribution to writing all parts of the report. His contribution was based on his joint book with Peter Kilby, Agriculture and Structural Transformation, and on the book Redesigning Rural Development: A Strategic Perspective, which he wrote in collaboration with William C. Clark. Dr. Johnston has requested that particular acknowledgement be given to Clark's contribution to some of the ideas presented here. The policy analysis perspective and the treatment of issues of organization and management drew heavily on chapters in their joint book (which was written mainly by Clark).

Dr. Johnston was, however, unable to collaborate with me in writing this book. His contribution is, thus, limited to writing the AID Report only, from which I extracted at length.

Dr. Lon Cesal was involved in all phases of the AID project, and made a major contribution to bringing the various facets of the study together into a final report. Also, he contributed to writing some sections of the AID Report which are not included in this book.

x

Special acknowledgement is due to I. G. Singh of the World Bank for generously providing the authors with his manuscript entitled <u>The Landless in Asia</u>. Many of his ideas have been incorporated in this book.

I wish to express special gratitude to Charles Hanrahan for his support and encouragement in initiating collaboration with AID, and to T. Kelley White for his administrative support throughout the preparation of the original report for AID. I also thank John Eriksson, Douglas Caton, and Don McLelland of AID for the support and spirit of cooperation in which the report was prepared.

Bandita Nayak provided excellent research assistance.

Gratitude is expressed to James Sayre who went above and beyond his official responsibilities to edit the manuscript. Arthur Dommen provided encouragement and moral support, and spent a considerable amount of time supervising the typing of the final drafts. Finally, Sterling Hawkins was extremely helpful with transferring the original manuscript from Lexitron to Wang, archiving and retrieving as needed.

Among those who provided typing support were Deloris Midgette, Bernadine Holland, Alma Young, Vicky Valentine, and Betty Acton.

Bela Mukhoti

I
Introduction

This book addresses the issue of agricultural development and the need for expanding employment opportunities and increasing the effective demand of the poor for food and other essential commodities in less developed countries (LDCs). Our basic premise is that an appropriate pattern of agricultural development can play a crucial role in attaining the multiple objectives of development. A strategy for developing countries emphasizing broad participation and employment-oriented production recognizes the need for a simultaneous emphasis on increased supply of agricultural output and expanded employment and income opportunities for low income families (thereby increasing the effective demand for agricultural output). Yet, such strategies tend to focus more on the supply than on the demand side of the equation. The analysis presented in this book focuses on both supply and demand, but emphasizes the importance of accelerating the rate of growth in effective demand of the poor in the LDCs via employment generation. The basic purpose of our study is to develop a conceptual framework that would help developing country governments and development assistance donors to identify policies and programs that are likely to be effective in increasing employment, effective demand, and agricultural output concurrently.

During the past decade, there has been a lively and inconclusive debate about the objectives and content of development strategies. In the 1950s and 1960s, development strategies focused primarily on increasing the growth of GNP by rapid industrialization and the transfer of "surplus" population from the traditional rural sector to the industrial and tertiary sectors. This approach emphasized the process of capital accumulation and the need to raise levels of savings and to supplement domestic resources with an inflow of external capital. For agriculture, however, there was also an emphasis, particularly in AID programs, on

building national institutions, especially in
agricultural research, education, and extension.
Support for India's agricultural universities during
this period was a notable example.

In terms of the growth of total GNP, the decades of
the 1950s, 1960s and 1970s could be judged to have been
rather successful for most developing countries. The
growth of per capita GNP, however, was much less satisfactory because of the persistence of rapid rates of
population growth in most of the lower-income developing countries. Moreover, mounting evidence indicated
that a large fraction of the population--especially in
rural areas--was benefiting very little from the
overall economic growth. As a result, a large and
growing number of families remained in a condition of
absolute poverty characterized by unemployment and
underemployment, widespread malnutrition, and high
rates of mortality and morbidity, particularly among
infants and small children.

One response to the shortcomings of previous
development efforts was to stress the importance of
expanding employment opportunities. Emphasis was also
placed on the reduction of poverty by meeting "basic
human needs." In the United States, amendments to the
Foreign Assistance Act mandated that an increasing
amount of aid be directed toward improving the
well-being of the poor majority. This focus on basic
human needs has served a useful purpose in emphasizing
that certain needs related to food, nutrition, and
health are indeed more "basic" than others. It also
underscored the importance of being concerned not only
with the growth of average GNP, but also with the
distribution of income gains among different segments
of the population and with the composition of the goods
and services produced and consumed.

There is no agreement, however, concerning the type
of development strategies that would be effective in
implementing a "basic human needs" approach. A major
thesis of this book is that emphasis on a dichotomy
between the goals of growth and of equitably satisfying
basic needs is unnecessary and unproductive. We argue
instead that development strategies must be concerned
with both the rate and the pattern of growth. In
particular, we emphasize the advantages of an
agricultural development strategy that is capable of
simultaneously achieving high rates of growth of
agricultural output and widespread increases in income
and in effective demand. These goals can be attained
by promoting the progressive modernization of the
small-scale family farms that predominate in a
developing country (Mukhoti, 1966, 1968, 1978, 1980). 1/

Such a strategy has significant economic advantages
in achieving sector-wide increases in agricultural
output at low cost. Based on capital-saving

technologies appropriate to the factor endowment of developing countries, it leads to fuller and more efficient utilization of the rural work force. In contrast, capital-intensive technologies lead to the displacement of labor from agriculture in a situation in which there are few opportunities for alternative employment, because of the very limited development of manufacturing and other nonfarm sectors. Moreover, emphasis on gradual but widespread increases in the use of divisible, yield-increasing inputs permits sizable increases in "total factor productivity"--that is, output per unit of all inputs--because technological innovations such as high-yield, fertilizer-responsive varieties enhance the productivity of the land and labor resources already committed to the agricultural sector. At the same time this type of agricultural strategy has significant social advantages. Expanding the opportunities for productive employment at a rate that exceeds the growth of the labor force seeking employment leads to a tightening of the labor supply/demand situation and to a steady and widespread increase in returns to labor. The resulting increases in incomes and in effective demand make possible the higher levels of food consumption needed to eliminate malnutrition and other serious manifestations of poverty.

Our emphasis in this book is on the effects of alternative development strategies on the rate and pattern of growth of agricultural output. But we recognize that those production-oriented policies and programs need to be supplemented by a selective strengthening of social service programs for education, health, and family planning. This is particularly true for infants and small children, because the high mortality and morbidity rates of that vulnerable group are a consequence of the two-way interactions between malnutrition and frequent bouts of diarrhea and other infectious diseases. In addition, efforts to slow the rapid growth of population, which accentuates the difficulty of achieving full employment and reducing poverty, can be facilitated by linking efforts to promote family planning with low-cost health programs that achieve broad coverage of a country's rural as well as urban population. Experience demonstrates that such programs can simultaneously improve the prospects for child survival and awareness of those improved prospects on the part of parents.

To simply assert the advantages of a broadly based, employment-oriented agricultural strategy is only a first step. An initial obstacle is that many development specialists and policymakers are skeptical of an agricultural strategy aimed at increasing the productivity of a country's small farmers. An exceptionally able Asian economist and former member of India's

Planning Commission has called attention to this problem when he emphasizes "that policymakers harbour serious doubts about a small-farm structure" and "regard it at best as an inefficient and transitional mode of production" (Krishna, 1979, p. 1). It is indeed transitional, but this transitional stage may span several decades. Because of the structural and demographic characteristics of the lower income developing countries, several decades will be needed for economic growth and structural transformation process of these predominantly agrarian economies to reduce their farm population and labor force. Only then the present trend toward an increasingly small average size of farm units will be reversed.

A priori reasoning is incapable of resolving the debate about the choice of strategy for agricultural development. The issues are so complex and the number of interacting variables so great that "intellectual cogitation" alone is not equal to the task of providing reliable guidance for the design and redesign of strategies for agricultural development. We have therefore supplemented a "thinking through" approach with an analysis and interpretation of past experience in Chapters II and III. This led to the identification of three alternative models of agricultural development: (1) the Dual-Size Structure Model (DSS model); (2) the Uniformly Small Farm Model (USF model); and (3) the Mixed Characteristics Model. The three models differ in terms of (a) equality or inequality in ownership and access to land and other assets; (b) macroeconomic policies affecting relative prices and access to resources; and (c) sectoral policies determining the type of technologies available to and adopted by farmers.

The historical experience of Japan and Taiwan is of special interest in demonstrating the feasibility and the desirability of pursuing a USF pattern of agricultural development. This pattern made it possible to achieve a rapid expansion of opportunities for productive employment and widespread increases in income and in effective demand. The development and diffusion of new technologies, investments in rural infrastructure, and actions related to other functional areas provided a basis for disperal strategies that enabled a large and increasing percentage of farm households to participate in the process of technological change and to benefit from increases in income.

The lessons to be derived from the historical experience of Japan and Taiwan take on added significance when one considers a theme that has dominated much of the development literature-- preoccupation with dualistic development models that have emphasized the existence of "surplus labor" in agriculture. These models often have been used to justify theories about the determination of agricultural wages and the incomes of farm households that have assumed or asserted that farm wages and earnings tend to be rigid. Much of this literature has emphasized an "institutional wage," a "subsistence wage," or even a "nutrition-based efficiency theory of wage," and as a consequence there has been a tendency to neglect the fundamental importance of factors influencing supply of labor and demand for labor. A recent critical review by Binswanger and Rosenzweig (1981) of employment, wages, tenure and other contractual arrangements in rural labor markets reaffirms the importance and considerable validity of "the principles of the supply-demand, competitive model," in spite of the institutional features that characterize rural labor markets in developing countries. They also stress that better understanding of the long-term changes in returns to labor calls for explicit study and analysis of "the reproductive and technological behavior that leads to the long-term evolution of supply and demand" (Binswanger and Rosenzweig, 1981, pp. 2, 55). We argue here that the relatively rapid growth in the demand for labor, the long-term increases in returns to labor, and the reductions in fertility that have been associated with agricultural development in Japan and Taiwan also tend to confirm the importance of a demand-supply framework. 2/

The development of high-yield, fertilizer-responsive varieties of rice and other major crops, combined with the expansion and improvement of irrigation and drainage, constituted the overwhelmingly important dispersal strategy in both Japan and Taiwan. Adoption of a gradually widening range of improved items of farm equipment also contributed to the growth of output. The farm equipment was simple and inexpensive and enhanced the productivity of labor rather than displacing it. Furthermore, demand for farm inputs, together with increased demand for consumer goods, provided an important stimulus to the decentralized growth of small- and medium-scale manufacturing units that employed relatively labor-intensive technologies.

During the post-World War II period, redistributive land reform programs in Japan and Taiwan reinforced the USF pattern of agricultural development and reduced the inequality in income distribution. It is important to recognize, however, that these countries were pursuing a USF pattern of agricultural development long before the redistribution of land ownership under the postwar land reforms. The considerable concentration of land in large ownership units, before the war was not reflected in the size distribution of operational or management units. Large landowners invariably rented out their land to tenants so that agricultural production was carried out by uniformly small units, although many of them were tenants or part-tenants rather than owner-cultivators. Because of the scarcity of land relative to the large number of farm households, the large landholders were able to demand rental payments equal to some 50 percent of the output produced by tenant cultivators. This resulted in a highly skewed pattern of income distribution. Nevertheless, tenants and landlords had a common interest in increasing productivity and output by adopting divisible, yield-increasing innovations appropriate to the labor-using, capital-saving technologies employed by the uniformly small farm units.

In contrast, the DSS pattern of agricultural development that prevails today in many developing countries is characterized by a concentration of agricultural land in a subsector of large and relatively capital-intensive farm enterprises. These large farms employ technologies which differ drastically from those employed by the great majority of farm units. Because of price distortions—underpricing of capital and foreign exchange—and other effects of macroeconomic and trade policies, the large-scale subsector has preferential access to resources. Moreover, since that subsector accounts for the bulk of commercialized production, the large farms are not subject to the severe constraints concerning cash income and purchasing power that characterize the agricultural sector in countries where the number of farm households is very large relative to the nonfarm population dependent on purchased food. Hence, large farm units are able to invest in labor-displacing mechanical equipment as well as in fertilizers and other yield-increasing inputs. Furthermore, this concentration of cash income in the large-scale subsector intensifies the constraint on purchasing power for the great majority of small farm units. In countries where land is scarce, the concentration of land in the large-scale subsector means that the size of the farm units in the small-scale subsector is even smaller than the small average size because the number of farm units in the small-scale subsector is so large relative to the total area of farm land.

Most of today's lower income developing countries confront a choice between the USF and DSS models because to a considerable extent these alternative patterns of agricultural development tend to be mutually exclusive. The intensified constraint on purchasing power that small farms face within a DSS pattern makes it exceedingly difficult to implement "dispersal strategies" in which government efforts, capital, and management skills are <u>dispersed</u> over essentially the entire agricultural sector. In addition, scarce resources of capital, foreign exchange, and trained manpower tend to be concentrated on "focus strategies" in which efforts are <u>focussed</u> on promoting rapid growth of output within large-scale farm enterprises, and capital as well as management resources are concentrated within that subsector. Focus strategies thus benefit the large subsector, and may also benefit a very limited number of small farms. But these latter strategies are so resource-intensive that they cannot be widely dispersed among the great majority of the farm population.

Many of the contemporary developing countries represent what we have referred to as a "mixed characteristics" model. This pattern of agricultural development involves a mixture of large and relatively capital-intensive farm enterprises coexisting with a much larger number of small-scale farms; but the large-scale subsector is not as dominant as in the DSS model.

We review the development experience of four countries--Kenya, Tanzania, Costa Rica, and Malaysia--nations which represent diverse conditions and illustrate very different examples of a mixed characteristics model. The experience of Malaysia is of particular interest. Owing to a virtually unique combination of factors, Malaysia is an exception to our generalization that a strong emphasis on focus strategies precludes the possibility of successfully implementing dispersal strategies. Our analysis of the special circumstances that enabled Malaysia to emphasize focus strategies in the development of a plantation sector, and at the same time to implement dispersal strategies that led to widespread increases in smallholder productivity and income, highlights the importance of certain characteristics of a developing country that are particularly relevant to the choice of an agricultural strategy.

Three factors--per capita income, the share of agriculture in the country's total labor force, and the nature of a country's resource endowment--are particularly important in determining the nature of a country's development problems and the strategic options that are feasible. The first two factors are particularly useful in defining a typology of developing countries because they are so highly correlated. A brief examination of data for 38 low-income, 52 middle-income, and 18 industrialized countries emphasizes, with few exceptions, that the low-income developing countries are also characterized by having a large share of their labor force in agriculture--77 percent in 1960 and still 72 percent in 1978. On average the share is much lower in the middle-income developing countries; and the decline in agriculture's share from 58 percent in 1960 to 45 percent in 1978 was considerably greater than the decline registered in the low-income countries. In the industrialized countries, the average share of agriculture in the labor force was only 17 percent in 1960, and by 1978 a mere 6 percent of the labor force was in agriculture. A brief review of trends in fertility, mortality, and in rates of natural increase lends support to the earlier statement that this structural characteristic of the low-income developing countries will continue to be a fundamentally important characteristic for many years. It is this characteristic, of course, that underscores the crucial importance of a pattern of agricultural development that fosters increases in farm productivity among the great majority of farm units, so as to generate expanded opportunities for productive employment in agriculture and widespread increases in income and in effective demand.

Even though a majority of developing countries may continue to correspond to our mixed characteristics model, we believe that empirical evidence and theoretical analysis both emphasize the importance of their approximating a USF pattern of development as closely as political and other constraints permit. Although we recognize the importance of political factors in shaping development strategies, we emphasize that they exert their influence in specific functional areas. We conclude that policies and investment programs in seven functional areas will largely determine the success of efforts to influence the rate and pattern of agricultural development: (1) asset distribution and access, (2) planning and policy analysis, (3) development and diffusion of new technology, (4) investments in rural infrastructure, (5) policies and programs related to marketing and storage, input supply, and credit, (6) rural industry and ancillary activities, (7) institutional development (improving organizational structures and managerial procedures).

In chapter IV we analyze each of these functional areas in order to assess the potential impact of policies and programs on employment generation and the increase of effective demand. The effects will, of course, be quite different in countries characterized by a USF pattern as compared with a DSS pattern. The adverse effects of a DSS pattern on employment expansion and on the growth of effective demand will be especially serious in the low-income, late-developing countries where the bulk of the population is still dependent on agriculture for employment and income. Differences in the availability of agricultural land and other features of a country's resource endowment also emphasize the location-specific nature of the problem of designing and implementing agricultural strategies.

In our discussion of the development and diffusion of new technologies, we emphasize that many of today's developing countries confront a more difficult task than Japan or Taiwan in evolving dispersal strategies. This is a consequence of their dependence on farming carried out under rainfed conditions rather than the relatively homogeneous and controlled conditions in Japan and Taiwan where irrigated agriculture predominates. Moreover, in many contemporary developing countries, especially in tropical Africa, the scope for expanding irrigation is limited, which means that agricultural research and extension programs must confront the special problems of increasing productivity and output among small farmers operating under rainfed conditions.

In chapter V, we address the problem of setting priorities for investment and development assistance. Priorities are identified in relation to the requirements for progress toward a USF pattern of development. This is justified because the USF model is so much more effective than the DSS model in providing employment for a rapidly expanding rural labor force, increasing agricultural output, and accelerating the growth in effective demand of the rural poor.

NOTES

1/ Complete citations to references are found at the end of the book.

2/ In addition to the general development literature, Binswanger and Rosenzweig give considerable attention to papers presented at a 1979 conference held in Hyderabad, India on "Adjustment Mechanisms in Rural Labor Markets in Developing Areas" sponsored by the Agricultural Development Council, the International Crops Research Institute for the Semi-Arid Tropics (ICRISAT), and the Ford Foundation. Chapters by Umemura, Tussing, Masui, Misawa, and Minami included in Agriculture and Economic Growth: Japan's Experience (Ohkawa, Johnston, and Kaneda, eds., 1969) are especially valuable as empirical and theoretical treatments of the evolution of the rural labor supply-demand situation in Japan.

II
Alternative Patterns of Agricultural Development: "DSS" and "USF" Development Models

Policy Analysis and the Importance of Learning from Experience

Agricultural and rural development are extraordinarily complex processes. Both the rate and pattern of change depend on a great many interacting variables: physical, economic, technological, demographic, institutional, and human. Concern with the growth rate of agricultural output is obviously critical; most developing countries must expand production by 2 to 3.5 percent annually merely to maintain per capita food supplies. Improvements in food consumption and nutrition also depend, however, on increases in the effective demand for food. This includes the "reservation demand" of farm households for subsistence consumption of their own production as well as the effective demand for purchased food by farm and, especially, nonfarm families. The pattern of agricultural development, that is the extent to which the entire farm population participates in increases in productivity and in agricultural income, has many other significant effects on rural well-being. The pattern has highly significant effects on overall economic growth and on structural transformation, the process whereby overwhelmingly agrarian economies are transformed into diversified and productive modern economies.

The fundamental proposition of this book concerns the economic and social advantages of broadly based, employment-oriented agricultural development. This proposition is much more persuasive as an empirical generalization supported by the analysis of historical experience than as a logical deduction. A policy analysis perspective emphasizes the limits of "intellectual cogitation" in thinking through solutions to problems of agricultural development and predicting their outcomes. Those limitations are especially severe because agricultural development is such a complex, ill-structured problem and the effects of government policies and programs on the rate and pattern of development depend on so many interacting

11

variables. 1/ Thus, the outcomes associated with
development efforts depend upon complex interactions
which cannot be controlled or predicted with much
precision. These include technical and economic
conditions, policies, institutions, and the responses
and performance of farmers, agricultural scientists,
administrators and field staff, private firms, and
other participants in the development process. Because
the essence of the challenge of agricultural development is to promote efficient, evolutionary change of a
complex, dynamic system, attempts to formulate agricultural plans on the basis of a subset of variables that
can be quantified satisfactorily will inevitably be
unsatisfactory because of the problems of "suboptimi-
zation." The need in a developing country is to design
(and redesign) an agricultural strategy--a mix of
policies and programs--which takes account of all of
the significant variables, including a number of
factors that are exceedingly difficult to quantify but
too important to ignore. It is also important to
emphasize that this is a continuing, adaptive process
which is guided by feedback derived from the experience
obtained in implementing programs and learning from
both successes and failures.

The 1950s and 1960s were characterized by
exaggerated expectations concerning the role of
economic planning based on an optimistic faith in man's
abilities to think through solutions to development
problems by intellectual cogitation. A statement by
India's Prime Minister Nehru epitomizes this optimistic
view of intellectual cogitation in its assertion that
planning and development "have become a sort of
mathematical problem which may be worked out scientif-
ically" (as quoted in Karanjia, 1960, p. 49). At the
opposite extreme is the approach to social problem
solving which Wildavsky (1979) refers to as "social
interaction," an approach which relies not on "thinking
through" but rather "acting out" solutions through
social processes--market-determined prices, bargaining,
voting, and other negotiated or trial-and-error
learning processes. In fact, intellectural cogitation
and social interaction each have important strengths
and weaknesses. Good policy analysis should emphasize
the complementary potential of the two approaches and
seek means of integrating them (Johnston and Clark,
1982, pp. 23-35).

The persistence of widespread and increasing rural
poverty in so many developing countries further underscores the fact that promoting agricultural development
is complex as well as intractable. The sobering
experience of the past 25 years also points to another
common pitfall in development planning: the tendency
to equate the feasible with the desirable. One version

of that pitfall is to assume that because a certain goal is so desirable, it must be feasible as well. In less developed countries, however, resources are scarce, needs are enormous, and there is never enough money, time, or trained manpower for all the important tasks demanding attention. Moreover, the fact that resources have a high opportunity cost means that the feasibility/desirability equation cuts both ways. That is, an apparently realistic penchant for sticking with those things that are demonstrably feasible may also be a pitfall because doing one thing almost always means not doing something else. All too often, opting for programs simply because they appear to be feasible is likely to preclude the search for other options which could have a much greater impact in reducing rural poverty and in furthering other development objectives.

A central thesis of this book is that historical experience, especially as illustrated by the patterns of agricultural development in Japan and Taiwan, provides a "model" of a strategy for agricultural development that is both feasible and desirable in simultaneously achieving high rates of growth of agricultural output and generating widespread increases in employment, income, and effective demand. The experience of Japan and Taiwan is especially significant in demonstrating that it is feasible to design and implement agricultural strategies that are effective in fostering rapid and widespread increases in productivity and output among small-scale farm units employing labor-using, capital-saving technologies. Their experience further demonstrates that such strategies have important economic advantages as a low-cost approach to expanding agricultural output and at the same time having significant social advantages in generating rapid expansion of opportunities for productive employment and widespread increases in effective demand for food and other essential goods and services. On the other hand, a country's agricultural strategy encouraging a dualistic pattern of agricultural development in which large farm units have preferential access to land and other resources largely precludes a broad-based, employment-oriented pattern of agricultural development.

Alternative Patterns of Agricultural Development

The Dual-Size Structure (DSS) Model

Agricultural development is characterized by a dual-size structure (DSS) in many developing countries especially in Latin America and Southern Africa. In this model, a relatively small number of atypically large and capital-intensive farm enterprises occupy a disproportionate share of the agricultural land. To cite one of the extreme examples, it is estimated that in Colombia the top 1 percentile of farmers account for 46 percent of the farmland and their holdings average over 1,000 hectares. Much of the land in these large farms is devoted to livestock rather than field crops; nonetheless, these large farmers account for a very large percentage of crop production as well as total agricultural output. Their share of commercial production of farm products is especially large. In contrast, the great majority of farm units are less than 10 hectares and account for a disproportionately small share of agricultural production. Since a large fraction of the production of small farms is for subsistence consumption, their share in commercial production is considerably smaller than their share in total output. In fact, production of coffee for export is the only significant source of cash income for Colombia's smallholders. Over half of the country's coffee farms are less than 5 acres. Although the small-scale farms account for much less than half of total coffee production, coffee farmers are in a privileged position among smallholders in Colombia and have considerably higher cash receipts than the typical small-scale farm unit (Johnston and Kilby, 1975, pp. 14-18).

The Gini coefficient, which varies from zero for a perfectly equal distribution to 1 for a completely uneven distribution (e.g., a situation where the top 10 percent of farm households accounts for all the farmland), is a convenient summary measure of the concentration of landownership. Not surprisingly, the Gini coefficient for Colombia is very high, although its coefficient of 0.87 is exceeded by the estimated Gini coefficients for Paraguay (0.94), Peru (0.94), Venezuela (0.93), and Chile (0.93). 2/ A number of other Latin American countries have similarly high Gini coefficients: 0.83 for Brazil and Guatemala, 0.80 for Nicaragua and the Dominican Republic, and at least 0.75 for Mexico. 3/ The estimated Gini coefficients for Pakistan and India are 0.63 and 0.58, respectively, based on FAO's Report on the 1960 Census of Agriculture. The concentration of land in those countries may have increased since that time.

The dual-size structure (DSS) model is also characterized by the use of drastically different technologies in the large-scale subsector as compared to those employed by the great majority of small farmers. Because the large farms tend to account for a substantial share of commercial sales, their cash receipts enable them to purchase expensive farm machines such as tractors as well as relatively large quantities of fertilizers and other inputs.

The opposite side of the coin, however, is that the great majority of small farmers are subject to an exceptionally severe purchasing-power constraint because the limited commercial market is largely preempted by the subsector of large farms. Consequently, they encounter great difficulty even in gradually expanding their use of divisible purchased inputs such as fertilizers needed to realize the high-yield potential of improved crop varieties. This cash income or purchasing power constraint is especially serious in "late-developing countries" where agriculture still accounts for a large percentage of the total population and labor force for reasons that we examine in Chapter III.

The foregoing difficulties of small farmers within a DSS pattern of agricultural development are an inevitable consequence of the concentration of land, commercial sales, and cash income among large-scale enterprises. Those disabilities are, however, usually intensified by the prevailing social environment and the concentration of political as well as economic power in the hands of the large farmers and their political allies and clients. The prevalence of economic policies that lead to the underpricing of tractors and other capital inputs and low interest rate exacerbate the consequences of the political power and influence of large farmers.

A low interest rate for loans obtained from cooperatives and other institutional sources represents an implicit subsidy for those who are fortunate enough to receive credit from those sources. Low interest rates also discourage saving and increase the demand for credit. Indeed, in inflationary situations, the official interest rates often represent a negative rate of interest in real terms and, therefore, an income transfer for those able to obtain loans. This combination of circumstances obviously gives rise to excess demand for the available supply of credit from institutional sources so that cooperatives and other institutional lenders must resort to administrative rationing of credit. Given the local power structure, it is the large farmers who obtain a major proportion of the institutional credit available; small farmers generally must obtain credit from money lenders and other informal sources or do without credit altogether.

Essentially the same circumstances frequently apply to the availability of fertilizers and other inputs. That is, government subsidies on those inputs, often adopted for the ostensible purpose of enabling low-income farmers to purchase fertilizers, give rise to an excess demand for the quantities available, again necessitating some form of administrative rationing.

The large farmers, because of their power and status, (including a situation in which they manage to "capture" control of the local cooperative) receive the bulk of the fertilizer and other subsidized inputs. Moreover, the fact that large subsidies on major farm inputs impose a substantial burden on the government budget often reinforces the effect of a general shortage of capital and of foreign exchange in limiting the total supply of those inputs.

The underpricing of tractors, often accentuated by trade policies, is another example of this problem. The combination of an overvalued exchange rate together with the granting of licenses for importing tractors and tractor-drawn equipment at zero or very low tariff rates enables large farmers to purchase labor-displacing equipment at artificially low prices.

Certain other consequences of the skewed distribution of political power associated with the DSS model should also be noted. Agricultural research, extension, and training programs are frequently biased toward the needs of large farmers. Examples of that bias are discussed in Chapters III and IV. Many developing countries have allocated considerable resources of money and manpower to training programs for tractor drivers and mechanics while research and development activity directed at identifying or developing well-designed, animal-drawn implements has either been sporadic and very limited or nonexistent. Extension field staff tend to devote most of their time and attention to meeting the needs of the large farmers; most small farmers rarely see an extension agent unless an extension program is structured to curb that tendency (Lowdermilk, 1972; Leonard, 1977).

An FAO analysis in Latin America, where this pattern is prevalent, shows a modern agricultural sector in which a small group of enterprises can take advantage of technological change and closer links with the rest of the economic system (The State of Food and Agriculture, 1978). The macroeconomic and other sectoral policies involving allocation of resources have favored modernization projects of a small group of influential and wealthy entrepreneurs. Application of modern technology and expansion of output have, therefore, tended to occur on a relatively small number of large or medium holdings. These farms generally have been the direct beneficiaries of much of the public investment, infrastructure, and other incentives in the form of credit, remunerative prices, protected markets, and extension.

This DSS type of modernization fits well into a framework of political conditions that tend to guarantee the stability of institutions and eliminate obstacles to rapid and unencumbered commercial activities. The report states that the agricultural entrepreneurs enter into various forms of agreement with financial groups, storage agencies, agro-industries, and the centers supplying modern technology. This modern subsector thus benefits from a much greater availability of capital goods, technological inputs, and other facilities. But increased food production in these countries, according to the same report, contrasts with "hunger and malnutrition among a large part of the population, increased agricultural incomes with spread of rural poverty, and progress in the application of modern technology with the persistence of primitive forms of agriculture" (The State of Food and Agriculture, 1978). The living conditions of the majority of the rural population, who depend basically on agriculture, continue to be depressed and precarious. Rural poverty has continued to spread despite substantial increases in average per capita agricultural output. Moreover, as the FAO report points out, "in many countries at least a quarter of the rural labor force is unemployed. Poverty, unemployment have caused rapid migration to urban areas, but the nonagricultural sectors have been unable to absorb this largely unskilled labor force, and poverty belts have grown up around the large urban centers." It is estimated that in Latin America between 1950 and 1976 over 40 million people migrated to urban areas. They represent almost half of the natural increase in the rural population. Moreover, about 46 million people in Latin America, or about 16 percent of the total population, were severely undernourished in 1972-74.

A final and important example concerns support for rural schooling. When large farmers dominate the local political process, the allocation of funds for public education generally receives a low priority. It is in the interest of large farmers to have a large supply of mainly unskilled labor available at low wage rates, and they are therefore often indifferent or hostile to using government resources to expand and strengthen education. This factor is presumably one of the principal reasons that the extent of education and literacy in some Latin American countries is below the level found in a number of Asian and African countries with much lower levels of average per capita income. 4/

The Uniformly Small Farm (USF) Model

An alternative pattern of agricultural development, best illustrated by the experience of Japan and Taiwan, is characterized by the progressive modernization of essentially all of a country's farm households.

Because the number of farm households in most developing countries is large relative to the total cultivated area, these farm units are inevitably small. This pattern is thus characterized by uniformly small farms, and we will refer to it as the USF model of agricultural development.

The term "uniformly small" is not to be construed narrowly as meaning equally small. Even though agricultural policies are designed to foster reasonably equal access to land, knowledge, credit and other resources, individual farmers will vary in the skill, intelligence, and energy that they apply in managing those resources. Access to land may be a result of landownership. In some situations, however, access is obtained by renting. Furthermore, we find that access to employment--farm or nonfarm--may provide satisfactory income-earning opportunities. In many developing countries, however, the job opportunities available to landless agricultural laborers are exceedingly precarious. Much of the most acute poverty is therefore found in the households of landless laborers.

The view that an effective and thoroughgoing land reform is a necessary precondition for a USF pattern of agricultural development is, superficially, reinforced by the fact that both Japan and Taiwan carried out remarkably successful land reform programs after World War II. However, the two countries were following a USF strategy long before the redistribution of landownership under the postwar land reforms. 5/

The considerable concentration of agricultural land in large ownership units was not reflected in the size distribution of operational or management units because the large landowners invariably rented out their land to tenants. Agricultural production was therefore based on uniformly small farms, although the farm units comprised roughly equal numbers of tenant, part-tenant, and owner-cultivator households. Sugarcane land in Taiwan represented the principal exception, although even for sugarcane much of the production was carried out on small farms which delivered their cane to a nearby sugar mill operated by a large plantation. The principal effect of the postwar land reforms was to give tenants and part-tenants title to the land that they cultivated and thereby to substantially reduce the inequality in rural income distribution. The economic rent associated with landownership now accrued to the individual cultivator rather than to the large land-lords collecting rent from tenants equal to some 50 percent of the output of the land they cultivated.

The most important consequence of the USF pattern was that the expansion of agricultural production was based on labor-using, capital-saving technologies

permitting widespread increases in farm productivity
and employment. In both Japan and Taiwan, the
development and diffusion of high-yield,
fertilizer-responsive varieties of rice was of central
importance. The large returns realized from those
divisible, yield-increasing innovations were also
associated with controlled irrigation. The development
and improvement of water control in Japan was a long,
evolutionary process. Taiwan's agriculture was
relatively undeveloped at the beginning of Japanese
rule in 1895. The expansion and improvement of
irrigation facilities was a major objective of the
Japanese colonial administration; investment by the
central government and the matching outlays by local
irrigation districts accounted for nearly 15 percent of
total capital investment in Taiwan during the 1920s.
Substantial investments were also made in extending
road and rail networks so that farmers throughout the
two countries had reasonably satisfactory transportation links with urban and industrial centers which
facilitated the marketing of their products and the
distribution of inputs.

There was naturally considerable individual
variation in the speed, skill, and energy with which
different farmers increased their productivity and
output. Such interfarm differentials tended to narrow
rapidly, however. The fact that the purchased inputs
required for the modernization process were divisible
and highly complementary to the onfarm resources of
labor and land meant that they could be adopted
universally in spite of the small size and limited cash
income of the uniformly small farm units. In Taiwan,
for example, close to 80 percent of all farms were
within 1 acre of the average size of about 2.5 acres.
Being highly divisible, the technologies were neutral
to scale. Indeed, in both Japan and Taiwan there was
an inverse correlation between farm size and crop
yields because the application of labor and fertilizer
on the small farm units was more intensive than on the
large farms. Moreover, because of the rate of technical change and its labor-using bias, increases in total
factor productivity (that is, output per unit of total
inputs) made a substantial contribution to the
impressive growth of agricultural production.

The fact that the USF patterns of agricultural
development in Japan and Taiwan depended so much on the
fuller, more efficient use of labor had very favorable
effects on employment expansion and increases in
effective demand. An especially signficant feature of
Taiwan's experience is that underemployment in
agriculture was reduced in spite of a substantial
increase in the size of the farm workforce in a

situation where there was only limited scope for enlarging the cultivated area. Following the intro- duction of public health measures by the Japanese colonial administration in Taiwan, mortality rates fell rapidly in rural as well as urban areas. The rate of natural increase in Taiwan, consequently, reached an annual rate of 2.2 percent as early as 1925-30. After a time lag of 15 years, this acceleration in population growth was followed by an increase in the rate of growth of both the total and farm labor force. From the turn of the century until about 1925, the cultivated area expanded more rapidly than the farm labor force so that there was some improvement in the land/man ratio. But over the extended period from 1911-15 to 1956-60, the cultivated land area in Taiwan increased by just over 25 percent, barely half the increase in the farm labor force. Between 1930 and 1960, the number of farm households increased much more rapidly than the cultivated area, resulting in a decline in the average farm size from 5 to 2.5 acres. Nevertheless, there is clear evidence of reduction in underemployment in agriculture.

The "flow" of labor inputs into agricultural production doubled between 1911-15 and 1956-60. This was twice the addition to the "stock" of farm labor because the average number of working days a year per worker increased by a third. This was facilitated by a wider adoption of multiple cropping with the result that the crop area nearly doubled even though the total area under cultivation went up by only 27 percent. The previously mentioned expansion of irrigation was obviously the critical factor in permitting the large increase in multiple cropping. The intensification of crop production between 1911-15 and 1956-60 was also associated with a thirteenfold increase in fertilizer consumption and a fivefold increase in all current inputs (Johnston and Kilby, 1975, p. 253). Technical innovations were land saving and farm outlays for labor-saving equipment were negligible until the 1960s when labor shortages finally began to emerge. Throughout this 50-year period, farm outlays for purchased inputs were concentrated overwhelmingly on divisible inputs of working capital that were complementary to the relatively abundant resource of farm labor. There was gradual improvement in the range and design of simple, inexpensive implements such as plows and harrows, row markers, and rotary weeders. But these items, important in easing seasonal bottle- necks and improving the timeliness and precision with which farming operations were carried out, did not displace labor. Investments in labor-saving equipment, notably power tillers, did not begin to become impor- tant until the process of structural transformation in

Taiwan reached a turning point in the 1960s and the absolute as well as the relative size of the labor force began to decline. In the first half of the 1960s, farm purchases of capital equipment represented about 25 percent of total outlays for farm inputs. But, prior to that period, purchases of current inputs were nearly 10 times as large as capital outlays (Johnston and Kilby, 1975, p. 318).

The USF pattern in Taiwan was associated with an increase in agricultural output at an average annual rate of 3.5 percent in both the prewar (1911-15 to 1936-40) and postwar periods (1951-55 to 1961-64). Moreover, a rapid increase in total factor productivity was the source of well over half of the increase in output. The fact that Taiwan's agricultural strategy was so efficient in its use of capital and other scarce resources also had important implications with respect to the net flow of resources to industry and other nonfarm sectors.

A distinctive feature of the USF strategies in Japan and Taiwan is that the positive interactions between agricultural and industrial development facilitated the concurrent growth of output and employment in agriculture and in manufacturing and other nonfarm sectors. In Japan, the growth of nonfarm employment between the 1880s and the 1920s was sufficiently rapid to permit a slight reduction in the absolute size of the agricultural labor force (from 15.5 to 14.3 million) and a substantial reduction in agriculture's share in the total labor force (from 76 to 52 percent). This was facilitated, however, by the fact that the demographic transition in Japan, as in Western Europe, was associated with a relatively moderate rate of population growth. The rate of increase in the total labor force in Japan was a little less than 1 percent.

In fact, Japan would have reached a structural transformation turning point characterized by a substantial reduction in its farm labor force during the 1920s if it had not been for the pursuit of economic policies during the interwar period which slowed the rate of increase in nonfarm employment. The deflationary policies necessitated by an unfortunate decision to maintain the yen at a level that was consistently overvalued between 1920 and 1932 were motivated by a desire to return to the Gold Standard at the prewar parity with the dollar and the pound sterling. Those deflationary policies had especially adverse effects on the growth of output and employment in the country's small- and medium-scale manufacturing firms which meant a marked slowing of the expansion of nonfarm employment opportunities. In the decades prior to the First World War and again during the period of rapid economic growth following World War II, expansion of output and employment in the relatively labor-intensive small- and medium-scale firms of Japan's "semi-modern" industrial sector played a major role in facilitating increases in the per capita income of the farm population by providing alternative employment opportunities, thereby permitting a reduction in the size of the population and labor force dependent on agriculture for income and employment.

The concurrent growth of output and employment in agriculture and industry in Japan and Taiwan was facilitated by a net flow of resources from agriculture to the more rapidly growing manufacturing and service sectors. This net outflow was exceptionally large in Taiwan and exceptionally well documented (Lee, 1971; Johnston and Kilby, 1975, chapter 8). From the point of view of the Taiwanese population, the net outflow of capital from agriculture was undoubtedly excessive since to a considerable extent the transfer of resources accrued to Japan. Consequently, the effects of the outflow of resources in accelerating the development of Taiwan's own nonfarm sector was not as great as implied by the size of the resource transfer which ranged between 20 and 30 percent of the value of agricultural output.

The rapid modernization and commercialization of Taiwan's agriculture which made possible the large net outflow of resources had positive as well as negative effects on Taiwan's farm population. As early as 1921-25, approximately 65 percent of total agricultural output in Taiwan was marketed in spite of the fact that nearly 70 percent of the country's labor force still depended on agriculture so that the domestic commercial market was very limited. The explanation for the high rate of commercialization of farm output is that well over half of the agricultural products that entered commercial channels were sold abroad, mainly rice and sugar exported to Japan. The substantial and effective investments made by the Japanese colonial administration in strengthening the physical and institutional infrastructure for agriculture were motivated by Japan's interest in fostering increased farm productivity and output in Taiwan in order to develop the colony as a supplier of imported sugar and rice for the Japanese home market. However, the establishment of agricultural experiment stations and research programs and the expansion of irrigation, transportation, and other infrastructure have been of immense and continuing value to the Taiwanese economy. More generally, the fact that the agricultural strategy pursued was based on the USF model which had already been so effective in Japan created conditions favorable for the very rapid and broadly based agricultural and industrial development achieved in Taiwan during the decades following World War II.

Equity and social justice were probably not high on the agenda on the Japanese rulers of Taiwan. Nevertheless, Taiwan's success in the postwar period in implementing a broad-based, employment-oriented agricultural strategy was to a considerable extent made possible by the progress made prior to World War II in implementing a USF model of agricultural development. Moreover, the gradual but widespread increases in farm incomes in Taiwan during that period generated a general growth of the effective demand for a widening range of simple, inexpensive items of farm equipment. This demand stimulated establishment and growth of rural-based, small-scale machine shops and other firms that fostered the growth and diffusion of technical and entrepreneurial skills. The growth of rural demand for relatively simple consumer goods was quantitatively more important, although the qualitative importance of the skills in metalworking acquired in producing all-metal plows, harrows, foot-pedal threshers, sweet potato slicers, and a host of other items of farm equipment may have been greater. However, no sharp distinction should be made between the two types of products because the same rural workshops often produced both consumer goods and farm implements. For both farmers and the entrepreneurs and skilled workers in rural-based manufacturing firms, this was a widespread, evolutionary process of upgrading skills and products based on learning-by-doing as well as a steady increase in cash incomes and capital formation. Many of the manufacturing firms were Japanese especially in the earlier period of Japanese rule. Both the learning and diffusion processes, however, benefited from the fact that the social and technological "distance" between the Japanese and Taiwanese was not nearly as great as between the "traditional sector" and the enclaves of "modern" manufacturing firms that have characterized former European colonies even after independence.

In summary, the USF model of agricultural development epitomized by Japan and Taiwan was characterized by rapid growth of employment opportunities within and outside the agricultural sector. Inasmuch as agriculture was essentially a "self-employment" sector dominated by small-scale farm units, most of the increase in onfarm employment reflected the increase in the "reservation demand" for family labor resulting from increases in productivity and output based on labor-using and land- and capital-saving technologies. In addition to the intensification of agricultural production with the adoption of divisible, yield-increasing innovations for rice and other crops, the income-earning opportunities of farm households were also augmented by the spread of ancillary activities.

A seventeenfold increase in the output of raw silk between the 1880s and the 1930s also made a notable contribution to the growth of farm cash incomes as well as to the expansion of foreign exchange earnings. This expansion continued through the 1930s in spite of a sharp decline in silk prices which began in 1925. This decline reflected the lack of alternative cutlets for the labor committed to sericulture. In addition, technical innovations, generated by research, led to remarkable increases in productivity and mitigated the adverse effects of the decline in silk prices. The production of mulberry leaves expanded much more rapidly than the increase in the area planted to mulberry trees. An enormous increase in cocoon production was facilitated by innovations which made it possible to raise an autumn as well as a spring crop, and the yield of raw silk per kilogram of cocoons nearly doubled (Johnston, 1962, pp. 229-30). In Taiwan during the post-World War II period, rapid expansion of the production of mushrooms and asparagus for export played an analogous role in expanding opportunities for productive employment of the agricultural labor force and in augmenting farm incomes. Finally, with the rapid and decentralized growth of manufacturing, members of farm families have been able to augment household income greatly by wages from nonfarm employment.

Until the absolute size of the farm labor force began to decline significantly during the 1950s in Japan and during the 1960s in Taiwan, the increases in per capita farm incomes were fairly modest; but, being widespread, the increases benefited virtually the entire farm population. Wage rates and returns to labor increased more rapidly with the tightening of the labor supply/demand situation as the demand for labor increased more rapidly than the growth of the farm labor force seeking employment opportunities. The growth of farm cash incomes led to especially rapid increases in outlays for farm inputs and purchases of manufactured consumer goods because of the high income elasticity of demand for those products. However, there was also a substantial increase in food consumption which, in the case of farm households, continued to be based on subsistence consumption as well as purchased food. Between 1953 and 1970, the Taiwan per capita availability of calories increased by 15 percent--from 2,300 to 2,700 calories per day--and the increases in protein and other nutrients were somewhat larger than the increase in energy intake. Inasmuch as the improvements in food consumption have

been so widespread, problems of malnutrition seem to
have been virtually eliminated (Galenson, 1979, pp.
436-37; Chiu, 1976).

The agricultural development experience of South
Korea has not been as well documented as the experience
of Japan and Taiwan. Japan also fostered a USF pattern
of agricultural development in Korea during the period
of colonial rule which began in 1910. Substantial
investments in institutional and physical infrastructure contributed to a broadly based, employment-
oriented pattern of agricultural development, and
increases in farm productivity and output were
encouraged in part in order to expand rice exports to
Japan. During the colonial period, much of the
farmland in Korea was owned by Japanese landlords, but
a land reform program in the postwar period created an
exceptionally uniform distribution of landownership.
The estimated Gini coefficient of 0.20 for the
distribution of farm land in Korea is remarkably low,
appreciably lower than the estimated coefficients of
0.41 and 0.40 for Japan and Taiwan, respectively (Berry
and Cline, 1979, p. 38).

The People's Republic of China (PRC) has relied
essentially on a similar USF pattern of agricultural
development in spite of the drastic differences between
China's Communist regime and the mixed economies of
Japan, Taiwan, and Korea which have relied primarily on
market mechanisms to guide the allocation of resources
and to determine the distribution of income. Since the
severe setbacks experienced in China in the late 1950s
and early 1960s, the fundamental unit of agricultural
production has been the production team made up of some
30 to 40 households. Agricultural decisionmaking has
for the most part been decentralized to this lowest
level of the commune structure, and the units have
apparently been small enough to maintain individual
incentives and to minimize problems of shirking and
poor performance. Thus, reasonably satisfactory results
have been obtained with production technologies which
have been labor-using and capital-saving. Likewise,
the increases in agricultural productivity and output
have been based to a large extent on improved varieties, fertilizer, and other divisible, yield-increasing
innovations together with very significant improvements
in irrigation and drainage.

This is, of course, in sharp contrast with the
Soviet Union where collectivized agriculture has been
characterized by a dual-size structure. Production in
collective enterprises in the Soviet Union has been
carried out in large-scale farm units employing large
tractors and tractor-drawn implements while at the same
time a substantial part of the country's agricultural
output has been produced on very small plots cultivated

by the family labor of kolkhoz (collective) households using extremely labor-intensive methods.

The overwhelming emphasis in the Soviet Union on large-scale manufacturing firms is another significant contrast with the PRC. China has given high priority to heavy industries based on large-scale, capital-intensive technologies. But as in the case of Japan and Taiwan, this has been paralleled by the decentralized growth of a "semi-modern" manufacturing sector producing relatively simple farm implements and consumer goods based on labor-intensive technologies. A number of mistakes were made in the earlier efforts to promote this rural-based industrialization. But, on balance, this approach has made a notable contribution to expanding the output of industrial and agricultural production and to providing additional opportunities for productive employment in rural areas (Perkins, 1977; Rawski, 1979).

In the countries of Sub-Saharan Africa, the expansion of agricultural production has taken place predominantly on small-scale holdings. This has, however, been essentially an "horizontal" expansion of production based on bringing additional land into cultivation. The technologies used have been very labor-intensive, relying mainly on human labor and the hoe and machete. Population growth has accounted for much of the increase in the supply of farm labor, but there has also been a significant increase in the rate of utilization of the "stock" of male labor as a result of a reduction in the time devoted to traditional activities such as hunting and fishing. The increase in male labor inputs in agriculture has been especially evident in the cultivation of new export crops such as cocoa, coffee, and cotton. To a large extent production of these new cash crops has been superimposed on the traditional systems of producing food crops, drawing upon the available "slack" represented by the underutilized resources of labor and land. The really important innovations were the economic innovations represented by the introduction of the new high-value crops. The expansion of cocoa in Ghana and of robusta coffee in Uganda and of food crops throughout tropical Africa are good examples of this largely spontaneous process of horizontal expansion of production. Cotton on the other hand, has relied considerably more on agricultural research which promoted the introduction of exotic varieties and later achieved fairly significant yield increases, especially by breeding for disease resistance (Anthony et al., 1979).

Although agricultural production in much of Sub-Saharan Africa has been based mainly on uniformly small units, it has for the most part continued to be a "resource-based" rather than a "science-based" agriculture. There have, of course, been many variations in this general pattern. In a number of countries, plantation production of palm oil and other export crops has been of considerable importance, and in a smaller number of countries large farms established by European settlers have been important in producing commercial crops. Some of the recent changes and future problems and prospects are illustrated in Chapter III by examining the experience of Kenya and Tanzania.

NOTES

1/ For a more complete presentation of a policy analysis approach to problems of agricultural and rural development, see Johnston and Clark, 1982.

2/ These and the other estimated Gini coefficients are from Berry and Cline (1979, pp. 38-39) and are based on Food and Agriculture Organization of the United Nations, <u>Report on the 1960 Census of Agriculture</u>, Vol. 5 (Rome, 1971).

3/ For Mexico, Berry and Cline report a Gini coefficient of 0.75 based on estimates which treat ejido land as equally distributed among all ejitarios (which it is not) and a coefficient of 0.95 as an unadjusted figure based on the <u>Report on the 1960 Census of Agriculture</u>, by the Food and Agriculture Organization of the United Nations.

4/ Guatemala is a prime example. Although its average per capita GNP in 1978 was some three to four times higher than that of the low-income developing countries of Sub-Saharan Africa, the percentage of its population of primary school age enrolled in school was only 45 percent in 1960 and still a modest 65 percent in 1977. In Tanzania, the increase was from only 25 percent in 1960 to 70 percent in 1977. In Colombia, with average per capita GNP comparable to Guatemala, 90 percent of the urban children aged 6 to 11 years were enrolled in school in 1974 but only 60 percent of rural children in that age group (World Bank, 1980, pp. 47, 154).

5/ Under the postwar land reform programs carried out in Japan, Taiwan, and South Korea, resident landlords were permitted to retain personally cultivated lands up to ceiling acreage, usually about 8 acres. Land held in excess of that ceiling for personally cultivated land, and all tenant-held land (except in Japan, resident landlords were permitted to retain 2.5 acres of tenant-held land), was taken over by the government with compensation. The land was then sold to former tenants, part-tenants, and the landless in small units to be cultivated by family labor.

III
Country Case Studies and
Typologies of Development Situations

<u>Focus and Dispersal Strategies:</u>
<u>Examples of a "Mixed Characteristics Model"</u>

The DSS and USF models examined in the preceding chapter represent polar extremes. In virtually all developing countries, we find some mixture of large and relatively capital-intensive farm enterprises coexisting with a much larger number of small-scale farms. Our definition of the two alternative models emphasizes, however, a contrast in government policies and their impact on the pattern of agricultural development. The essence of the DSS model is that efforts are focused on promoting rapid growth of output within large-scale farm enterprises, and resources of capital and management are therefore concentrated within that subsector. In contrast, the essence of the USF model is that policies and programs are designed to promote the widespread, progressive modernization of a large and increasing percentage of a country's small-scale farm units. The emphasis on achieving broad coverage means that government efforts and resources of capital and management skills must be dispersed over essentially the entire agricultural sector.

In the design of agricultural development projects and programs, policymakers in national governments and in foreign aid agencies such as the World Bank or the U.S. Agency for International Development face a choice between what we will refer to as focus and dispersal strategies. 1/

If governments emphasize focus strategies in which resources are concentrated within a subsector of large-scale farm enterprises or within limited areas, the outcome will conform with our DSS model. Conversely, a USF pattern of agricultural development requires that government policy emphasize the design and implementation of dispersal strategies that will be effective in fostering widespread increases in productivity and output among a large and growing number of small-scale farm units. A key characteristic of these small-scale units is that increases in productivity and output are based on labor-using, capital-saving technologies which expand employment opportunities for family labor, and in many instances, hired labor as well. We argue later that there are important trade-offs involved in attempting to implement simultaneously focus and dispersal strategies. In fact, if there is a strong emphasis on a focus approach leading to a DSS pattern of development, this will tend to preclude the possibility of successfully implementing dispersal strategies consistent with the USF model.

The nature and seriousness of competitive tradeoffs between focus and dispersal strategies depend on the characteristics of particular countries. Those characteristics are stressed in defining the typologies of developing countries presented in the next section. The four country case studies presented in this section, however, illustrate the influence of some of those characteristics. Although our emphasis in this section is on the relative importance of focus and dispersal strategies, it is stressed later that a country's pattern of agricultural development will also be influenced strongly by certain macro policies, for example, policies affecting the relative prices of capital and labor.

Kenya

Agricultural development in Kenya in the decade prior to its independence in 1963 was one of the more extreme examples of a DSS model among the countries of Sub-Saharan Africa. The top 1 percent of farm units accounted for just over 50 percent of all farmland. The estimated Gini coefficient of land concentration was 0.82, comparable to the situation in Latin America. The colonial government's focus strategies for agricultural development encouraged plantation production of coffee and tea on large European estates and the promotion of large-scale arable farming in the so-called White Highlands. The growing concentration of the African population in native reserves was one of the major factors generating support for the Mau Mau movement against colonial rule.

The British Government's response to that challenge was not limited to military action against the Mau Mau. In 1954, the colonial government launched for the first time a major dispersal strategy aimed at promoting agricultural development among African smallholders. That strategy, widely known as the "Swynnerton Plan," was a multifaceted approach to raising farm productivity and incomes. These efforts constituted a set of dispersal strategies, some of which were implemented concurrently and focused on specific crops in different areas depending upon local ecological conditions and some of which were introduced sequentially. The major efforts and successes of this scheme were realized in "high-potential" areas such as the Central Province lying to the north of Nairobi and in Kisii District, another high-altitude area with reliable rainfall and good soils. An important feature of this new approach was the linking of efforts to improve farm practices on consolidated holdings with a new policy of promoting the planting of high-value crops such as coffee, tea, or pyrethrum by African smallholders.

The success of that dispersal strategy was facilitated greatly by the availability of technical knowledge about these crops accumulated by Kenya's agricultural research stations and European farmers. Moreover, there was a strong pent-up demand by African farmers to grow these profitable crops, production of which had long been restricted to European farmers. Because of the Mau Mau emergency, there had been a considerable strengthening of field staff in the areas where the dispersal strategy was being implemented. Substantial funds were made available for expanding nurseries to provide planting material, to strengthen agricultural extension work, to make loans and grants to support production and processing, and to help develop cooperatives and other marketing organizations. The results can be illustrated by the dramatic increase in smallholder coffee production. From a level of less than 1,000 tons in 1954-55, production of Arabica coffee by smallholders rose to 4,600 tons in 1959-60, 16,600 tons in 1964, and 39,300 tons in 1974. By 1964, smallholder production already accounted for 41 percent of Kenya's total coffee production. By 1974, the smallholder sector accounted for 70 percent of total output (Johnston, 1964, p. 172; Senga, 1978, p. 81).

The dispersal strategy for smallholder tea illustrates another dimension of the overall effort and is of particular interest. Tea had long been regarded as a "plantation crop" par excellence because of the need for close coordination between tea growing, plucking of the tea leaves, timely collection of the leaves, their transport to tea factories, and processing which requires a substantial investment in plant and equipment. By the institutional innovation of creating a quasi-governmental organization, the Kenya Tea Development Authority (KTDA), it was possible for smallholders to rapidly expand tea production, an activity that lends itself to labor-intensive, small-scale production. The KTDA performed the support functions of transportation and processing which require coordination and which are characterized by scale economies.

Initial emphasis in Kenya was on introduction of high-value crops destined for export. This was soon followed by an effective dispersal strategy for promoting smallholder production of hybrid maize. It was not until 1955 that a research officer was appointed to work exclusively on maize at the National Agricultural Research Station at Kitale, the station assigned principal responsibility for maize breeding. The small but competent core staff working on maize breeding was augmented effectively at critical periods by specialists provided under well-conceived technical assistance programs. Kenya's maize program was also able to obtain seed of promising varieties from high-altitude locations in Mexico and Colombia because of cooperation with CIMMYT (Centro Internacional de Mejoramiento de Maiz y Trigo) and other international programs. The success achieved in promoting the spread of the high-yield varieties among smallholders was remarkable, especially considering the need for annual replacement of the seed to maintain high yields with hybrid maize. In 1973, some 90 percent of the small-holders were growing hybrid maize in two of three survey areas in western Kenya. Medium maturity hybrid varieties adapted to growing conditions in Kenya's Central Province did not become available until 1970. The rate of diffusion in that region, however, was apparently even more rapid than in western Kenya: 55 percent of farmers in Central Province were already planting hybrid seed in 1974 (Gerhardt, 1975; Hessel-mark, 1975).

Experience in the higher elevations of the Kisii highlands provides an especially interesting illustration of the sequential and cumulative nature of the progressive modernization of smallholder agriculture in areas where the farming environment was well suited to the dispersal strategies. Prior to the 1950s, the development of agriculture in Kisii District was typical of the "horizontal expansion" process referred to in the preceding section. The opening up of the Kisii highlands to agricultural settlement had already led to a fairly rapid expansion of cultivation and production based on essentially traditional farming methods. Acceleration of population growth provided a strong stimulus to the rapid colonization of the Kisii highlands in the 1940s and 1950s; the population of Kisii District rose from 225,000 to 519,000 between 1948 and 1962, according to census estimates. Although farming methods and crops remained essentially traditional until the 1950s, some technical changes had been introduced. Imported English hoes began to spread rapidly shortly before World War I. And, in the 1920s, the distribution of seed of an improved maize variety contributed to increases in productivity and output.

Introduction of pyrethrum in 1952 under the Swynnerton Plan initiated a period of impressive agricultural progress. Earnings from pyrethrum provided the capital which made investment in small-holder tea possible. Production of high-value tea expanded rapidly from 1957. The KTDA played a critical role in providing the supporting services essential to the success to smallholder tea production, including the construction of "tea roads" for transporting leaf to factories for processing. By 1963, conditions were favorable for the launching of another dispersal strategy: the introduction of highly productive "grade" cattle for milk production. These cattle are a cross between exotic dairy breeds from Europe and indigenous animals. Although the crossing with the local zebu cattle provides some immunity from local disease and pest problems, the grade cattle must be protected in enclosures, sprayed for ticks, and properly fed. This requires more complex management, including rotational grazing, regular purchases of feed concentrates, and the use of artificial insemination. By the late 1960s varieties of hybrid maize adapted to conditions in the Kisii highlands became available and began to spread rapidly. In this area, the maize was mainly grown for home consumption rather than as a commercial crop because tea and other alternatives were more profitable in spite of the high-yield potential of the hybrid varieties. This sequence of innovations was very efficient in fostering cumulative change. Each successive innovation produced increased cash income and capital for later innovations and also predisposed farmers to accept other and more demanding technologies (Uchendu and Anthony, 1975; Anthony et al., 1979, pp. 183-85).

Although notable success was achieved in Kenya in implementing a number of dispersal strategies, the pattern of agricultural development has fallen short of the USF model for a number of reasons. The principal deficiency is that the dispersal strategies have not encompassed the entire agriculture sector and have only been suitable for the ecological conditions found in Kenya's high-potential farming area. This is clearly evident in the experience in diffusing high-yield hybrid maize. By 1973, some 90 percent of the small-holders had adopted hybrid maize in two of three survey areas in western Kenya. In contrast, only about 15 percent of farmers in the third zone were planting hybrid varieties. The explanation seems clear. This is a lower altitude zone which has somewhat lower and less reliable rainfall. As a result, the yield advantage of the hybrid varieties was much less and also more uncertain than in the other two zones.

Major agricultural regions in Kenya have experienced very little if any improvement in agricultural productivity and income during the period since independence. This is true of much of the Western Province and of the Eastern and Coast provinces. It has also been true of most of the pastoral groups which for the most part subsist from grazing livestock over large but agriculturally inhospitable areas.

The failure to design and implement dispersal strategies adapted to the entire agricultural sector, and especially the environmental conditions prevailing in Kenya's "medium potential" (semiarid) and "low potential" (arid) zones, can be attributed primarily to the lack of effective research directed at the problems and development possibilities of those areas. As a result, a large and growing percentage of Kenya's rural population has not benefited from improved income-earning opportunities, and many appear to have experienced a deterioration in their well-being. These problems are epitomized by developments during the past two decades in the semiarid areas of eastern Kenya, especially parts of Kitui, Machakos, and Embu districts.

Growing land scarcity in the high-potential areas has created pressures which have induced a substantial percentage of the population in those areas to migrate to areas of low and erratic rainfall where crop failure and famine are increasingly serious problems. Progress has been made in developing a short-maturity, drought-tolerant variety of maize which has a yield advantage over local varieties, especially in seasons of below normal rainfall. Research has not yet been sufficient to provide the basis for viable farming systems adapted to these semiarid conditions. The spread of the short-maturity maize may in fact have accentuated famine problems because it has encouraged the substitution of maize for sorghum and millet, the traditional drought-resistant cereals, better suited to the marginal rainfall of these semiarid areas. Other factors, including the greater vulnerability of sorghum and millet to bird damage, have also contributed to the shift. The most fundamental factor, however, has probably been the failure to develop and diffuse high-yielding varieties of sorghum and millet. Research needed to evolve more productive farming systems for these areas will also require research on farm equipment and tillage innovations to permit the adoption of moisture- and soil-conserving techniques together with other changes in farming practices. It is only in the last six years that Kenya has made a serious beginning with research to identify farm equipment suited to the needs of small farmers with limited cash income and purchasing power. This is a problem to which we will return in the next section because it is relevant to the task of increasing farm productivity under rainfed conditions in many of today's developing countries.

Rapid population growth is perhaps the most fundamental factor responsible for the failure of the rural population to benefit from increases in farm productivity and income. Kenya's rate of natural increase is approximately 3.9 percent annually, and the growth rate of the working age population is approaching that rapid rate of increase. Kenya's manufacturing and other non-farm sectors are still very small, and much of the industrial investment has been capital-intensive. As a result, the growth of the farm population and labor force has also been extremely rapid.

Rural-to-rural migration to marginal farming areas has been an important, if unreliable, safety valve for this growth of the rural population and labor force. For example, census figures show a 14 percent annual increase of population in the semiarid areas of Machakos District between 1963 and 1969, primarily as a result of the influx of population from more productive but densely populated farming areas. This rapid growth of population in farming areas with low and erratic rainfall has continued. One consequence has been to magnify the demands for famine relief on the frequent occasions when the level or distribution of rainfall is more unsatisfactory than usual. This hardship of the households affected is also accompanied by soil erosion aggravated by the cutting of trees and shrubs for charcoal and firewood. Charcoal-making is an especially important source of income for purchasing food in seasons when crops fail.

Settlement schemes and the spontaneous movement of population into the former White Highlands has provided another safety valve for the growing population pressure in rural areas. The land available in that region, however, has been limited because many of the former European farms are now being farmed as large, mechanized units by African entrepreneurs. There will probably be continuing and increasing pressure to subdivide those large holdings. However, wheat is still a major crop in much of that region; the technical knowledge and experience for wheat production in Kenya is limited to large, mechanized farm units.

Projections of Kenya's population and labor force growth indicate that the problems arising from population pressure in rural areas will intensify. Kenya's development plan for 1979-83 called for family planning to start a decline in fertility among the rural population, accounting for about 80 percent of the total population. Declining fertility should cut the increase in population between 1978 and 2000 to "only"

about 90 percent, compared with more than 130 percent
with constant fertility. But, there is a time lag
before a reduction in birth rate generates a reduction
in the labor force rate of increase. Therefore, the
118-percent projected increase in the working age
population with declining fertility is only a little
less than the 125-percent increase anticipated with a
constant birth rate (Kenya, 1979, p. 63). The employment implications of rapid growth of Kenya's population
and labor force become more dramatic if the time
horizon is extended. A set of projections tracing the
growth of population and labor force between 1969 and
2024 is particularly valuable because the projections
are broken down by rural-urban location and by age. On
the basis of the most likely scenario, Kenya's rural
labor force would decline from 87 percent of the total
labor force in 1969 to 65 percent of the total in
2024. The rural labor force would increase 4 times and
the economically active population in urban areas would
increase 16 times (Shah and Willekens, 1978, pp. 29,
38).

This demographic outlook underscores the importance
for Kenya to realize a USF pattern of agricultural
development. Expanding employment opportunities for a
rapidly growing labor force is so challenging that
vigorous and well-conceived efforts are needed to
expand income-earning opportunities in agriculture and
to accelerate growth of nonfarm employment opportunities. Slow progress in bringing about a decline in
fertility will, of course, make the task of expanding
opportunities for more productive employment and raising incomes and effective demand even more formidable.

Tanzania

This account of the Tanzanian version of the "mixed
characteristics" model will be even more selective and
incomplete than the treatment of Kenya's experience.
The approach to agricultural development in Tanzania
has been influenced strongly by President Nyerere's
concept of _ujamaa_ and African socialism particularly
since the Arusha Declaration of 1967. Prior to that
declaration there were many similarities between the
experience in Tanzania and the development of
smallholder agriculture in Kenya.

Farming by European settlers in Tanzania
(Tanganyika until the union with Zanzibar in 1964) was
much less important than in Kenya. Partly for that
reason, smallholder production of coffee was not
restricted and began much earlier than in Kenya. In
addition to Arabica coffee production in the high-
altitude areas of Kilimanjaro, there was expansion of
Robusta coffee west of Lake Victoria and impressive

growth of cotton production in Sukumaland. Production
of tobacco and several other export crops increased
rapidly during the 1950s and 1960s. The fact that the
dynamic crop sectors were all based on production for
export is not surprising because the domestic commercial market for farm products in Tanzania is even
smaller than in Kenya. Production of maize and other
food crops expanded in pace with the growth of
population until the mid-1970s when there was sharply
increased reliance on imports.

The rapid expansion of cotton production in Sukumaland in the 1950s and 1960s offers a good illustration
of the essentially "horizontal" expansion of smallholder production stressed earlier. Cotton was first
introduced during the period of German colonial
administration and was subsequently promoted by the
British administration. During the period 1935-44,
annual production averaged about 33,500 bales (of 400
pounds net weight). Production in 1962 rose to 200,000
bales reaching a peak of 400,000 bales in 1966 (Uchendu
and Anthony, 1974, p. 14). Plant breeders, producing
varieties with resistance to the cotton jassid and to
bacterial blight, played an important role in raising
yields. But expansion of the cultivated area was the
dominant factor leading to increases in output. A
thirteenfold expansion in cotton production in Geita
District is a striking example of the expansion process
which began to accelerate in the 1930s. Removal of
obstacles to settlement in the district--tse-tse fly
and the lack of water for human and livestock needs--
led to rapid immigration from parts of Sukumaland
experiencing population pressure. Population in Geita
District increased about six times between the mid-
1930s and the late 1960s. There was also a significant
increase in commercial production of cassava in Geita
for sale to other areas in Sukumaland where land scarcity and specialization in cotton production was generating rural as well as urban demand for purchased food.

On the eve of Tanzanian independence, a World Bank
mission to Tanganyika placed major emphasis on a
"transformation" approach for "planned and supervised
settlement of areas which are at present uninhabited or
thinly inhabited" (World Bank, 1961, p. 129). The
mission noted that a transformation approach could also
be carried out "through intensive campaigns in settled
areas, involving a variety of coordinated measures..."
The mission's report argues, however, that settlement
was the more promising method of implementing a
transformation approach: "When people move to new
areas, they are likely to be more prepared for and
receptive of change than when they remain in their
familiar surroundings" (p. 131). An important contrast
between Tanzania and Kenya is that Tanzania still has

extensive areas with good soils and adequate rainfall for further expansion of its cultivated area. With a rate of natural increase estimated at just over 3 percent per year, population pressures on the land in the earlier developed areas such as Kilimanjaro and most of Sukumaland rose. Until recently, however, large regions of the country with good agricultural potential were almost totally without transportation links. The building of the railroad and highway to link the port of Dar es Salaam with Zambia has been an important step toward opening up some of this land to settlement. But the railroad and highway were only a beginning; substantial investments in secondary roads and other infrastructure will be required to take advantage of the potential.

The "transformation" approach recommended in the World Bank report contrasts with an "improvement" approach which aims at more gradual change of the existing farming systems. These concepts parallel the distinction we have made between "focus" and "dispersal" strategies. We consider it important, however, that the concept of a "dispersal" strategy carries a clear implication that agricultural research has generated feasible and profitable innovations so that extension and related programs have something worthwhile to "disperse." All too often the "improvement" program promoted by agricultural officers in Tanzania and other African countries were derived from rather vague concepts of good farming and were not based on an adequate understanding of the conditions and constraints faced by smallholders. There was, for example, an almost universal emphasis on promoting production of crops in pure stand. Recently, agricultural scientists have recognized that the common traditional practice of mixed cropping often has significant economic advantages.

The new government of Tanzania in 1962 established a Village Settlement Agency to "transform" peasant agriculture along the lines recommended by the World Bank. Some 30 settlements were set up in various parts of the country with settlers recruited from areas of land shortage. By the use of tractors, substantial areas were put into cultivation, but the settlers could not cope with the problems of weeding and other subsequent operations. There were also problems with the maintenance and repair of equipment. The basic problem, however, was that production costs, which in the case of the "block farms" for cotton in Sukumaland included the use of aircraft for spraying, were excessive. In virtually all of the schemes, debt repayment was poor and most schemes soon approached bankruptcy.

By 1966, the experiment was declared a failure, in

part because of its emphasis on "premature mechanization." It was, therefore, "decided that, instead of establishing highly capitalized schemes and moving people to them emphasis shall be on modernizing existing traditional villages..." (statement by second vice president Rashidi Kawawa in April 1966 as quoted in Hyden, 1980, p. 74). The strong emphasis on the village settlement schemes and block farms had an adverse impact on extension efforts directed at an "improvement" approach because of the concentration of staff and government funds on the transformation approach. Nevertheless, farmers in a number of regions were making considerable progress in raising their productivity and output.

The Arusha Declaration of 1967 marked the beginning of a new phase of agricultural policy in Tanzania. In his essay on "Socialism and Rural Development," President Nyerere had emphasized ujamaa (literally family-hood) as a basis for the country's agricultural strategy. 2/ The goal was to establish communal village production units. The most notable increases in agricultural production during the 1950s and 1960s had been realized by "progressive farmers" who often received preferential treatment by extension staff and in other government programs. With the stronger emphasis on socialism following the Arusha Declaration, the emergence of these "petty-capitalist farmers" was regarded as a threat to "the growth of a socialist Tanzania in which all citizens could be assured of human dignity and equality..." As Nyerere's essay noted, Tanzania's ujamaa approach seems to have been inspired by the People's Republic of China (PRC), which Nyerere visited in 1965, and some local initiatives toward communal production. The most important of the latter was the Ruvuma Development Association (RDA) in Ruvuma Region. This was an attempt to improve existing practices and to promote increased cooperation quite independent of government influence. The RDA chairman provided dynamic leadership but worked closely with village leaders without imposing his ideas (Hyden, 1980, p. 76).

For some years the government relied on a voluntary approach in its effort to spread the ujamaa concept, although local officials were under considerable pressure to demonstrate results. In 1967 a regional development fund was created exclusively for the advancement of ujamaa villages. Tractors and other inputs were made available to selected villages as an inducement for the development of communal production. This, as Hyden points out, was essentially a form of patronage (Hyden, 1980, pp. 108-11). It was also a very inefficient form of a focus approach. In addition to problems of maintenance and repair, the tractors were often used inefficiently to cultivate communal fields that were not properly tended because villagers gave priority to their individual plots. The tendency to shirk on communal production is an ubiquitous and important problem which we will examine in some detail in later sections.

The voluntary approach to promoting ujamaa villages made slow progress. In November 1973, President Nyerere announced that by the end of 1976 all Tanzanians would have to live in villages. Although villagization was made compulsory, communal cultivation was not. Most villages set up during this enormous resettlement operation are "development villages" rather than ujamaa villages, i.e., characterized by a substantial amount of communal farming.

The launching of the program of compulsory villagization coincided with the beginning of a severe drought, making reasons for the subsequent decline in food production unclear. Weather conditions were undoubtedly important, but many would concur with the explanation advanced by Lofchie (1978, p. 452): "There is compelling reason to believe that the programme of collective villagization was the major cause of a crisis in agricultural production of calamitous proportions." His emphasis is on the resistance of peasants to an unpopular program, whereas others stress the poor timing and the administrative ineptitude displayed in carrying out the compulsory villagization program. In contrast, Samoff (1981, pp. 298-99) argues that the villagization program had little impact on the 1973-75 food crisis.

Official statements have consistently emphasized that Tanzania's development efforts are aimed at improving the well-being of the overwhelming majority of the country's population which continues to live in the countryside. Furthermore, Tanzania is often cited as an example of a country giving priority to the basic needs of its people because of its egalitarian approach to development. For example, a monograph on agricultural mechanization by a team of West German scholars emphasizes that many developing countries have encouraged labor-displacing mechanization with adverse effects on the mass of the farm population. The authors assert that "Tanzania is one of the few countries that adopted a policy of both selective and appropriate mechanization..." (Bodenstedt et al., 1977, p. 112). This assertion is supported by a reference to Tanzania's second 5-year plan (1969-1974) which called for the widespread introduction of ox cultivation, especially through the establishment of ox-training centers. An FAO mission on agricultural mechanization reported, however, that members found it extremely difficult to find any concrete evidence in the regions of a concerted effort to promote wider and more intensive use of animal power. The program of establishing ox-training centers, mentioned in the 5-year plan, appeared to be virtually nonoperative (FAO, 1975, p. 9).

This conclusion is supported by later reviews of Tanazania's activities related to "appropriate mechanization" (Beeney, 1975; ILO, 1978). Beeney's FAO/UNDP study points out some reasons why TAMTU, Tanzania's Agricultural Machinery and Testing Unit located near Arusha, was so inefficient in promoting wider, more efficient use of animal-powered equipment adapted to the requirements of small-scale farm units. He notes that agricultural engineers working in isolation cannot identify farm equipment and tillage innovations that will be appropriate and profitable within specific farming systems with the variety of soil and other agroclimatic conditions which determine the nature of those systems.

If a strong emphasis on a focus strategy jeopardizes prospects for dispersal strategies, Tanzania's current emphasis on state farms is a cause for concern. J. H. J. Maeda (1981, p. 137), Director of the Institute of Development Studies at the University of Dar es Salaam, recently reiterated Tanzania's policy of emphasizing "people-centered agrarian development." He states: "Since some 95 percent of Tanzania's total population is comprised of rural peasants who earn their livelihood through subsistence farming, it is quite natural that our development policies should concentrate on the needs of the subsistence farmer." Maeda (p. 143) cites "better use of available resources to increase production as producers take advantage of economies of scale in production, purchasing, marketing, mechanization, and the provision of required social and economic infrastructure" as the first of four considerations which justify the 1973 decision to carry out a compulsory program of villagization. Clearly, grouping scattered farm households into villages facilitates the task of making available health and school facilities and other social services. More questionable is the emphasis on economies of scale in production and mechanization. Furthermore, the concentration of rural households in large villages creates problems for efficient and sustained agricultural production. Unless the villagers are prepared to walk long distances to their fields, there is a tendency to shorten fallow periods. This accentuates problems of soil erosion and loss of fertility unless traditional farming systems are modified drastically and extensive use is made of purchased inputs, particularly fertilizers.

There is a strong Marxist tradition deriving from Marx, Kautsky, and Lenin, which emphasizes economies of scale in agriculture. This belief seems clearly to have influenced the Soviet decision of 1928 to reorganize agricultural production on the basis of collective farms dependent on Machine Tractor Stations (Johnson and Kilby, 1975, pp. 277-88). The earlier optimism over communal production in Tanzania seems based on expected increases of productivity and output from economies of scale. The decision to establish rural communes in the PRC has the same root. The pragmatic leadership in the PRC revised that policy in the early 1960s and gave priority to production teams as the principal unit of farm decisionmaking and management. The production teams appear to be small enough to minimize the problems of labor shirking and poor performance, especially serious when production is based on labor-using, capital-saving technologies.

Tanzania's current emphasis on large, mechanized state farms is probably a reaction to the difficulties encountered in increasing agricultural production and sales by peasant farmers in the village sector. Thus, Lofchie (1978, p. 473) reports indications that "the country's highest ranking officials are pessimistic about the possibility of improving agricultural production in the villages and are, instead, turning increasingly to large-scale state and private farms to accomplish that purpose." A number of state farms were set up shortly after independence, but they were concentrated on sisal and wheat production on large farms previously operated by foreign-owned plantations or European settlers. In 1976, however, a decision was adopted to set up six state farms for large-scale, mechanized production of maize. In addition to the importance attached to scale economies, this decision was apparently motivated by the need to supply urban centers after the state marketing agencies experienced problems in purchasing food from scattered peasant producers (Hyden, 1980, pp. 141-42). This emphasis on setting up state farms for food production apparently continues. Moreover, government officials do not recognize that concentration of scarce manpower and money in expanding the state farm sector will inevitably deprive the village sector of the resources and support services that are needed to increase productivity and output (ILO, 1978). Therefore, the pessimistic view of improving village agricultural production, reported by Lofchie, seems likely to become a self-fulfilling prophecy. This will, of course, increase the likelihood that a DSS pattern of agricultural develoment will prevail in spite of the emphasis of government policy on basic needs and "people-centered agrarian development."

Costa Rica

Every country represents a unique combination of physical environment, historical influences, and socioeconomic factors. In Costa Rica, a small and relatively prosperous developing country with a population of 2.2 million, the unique features are exceptionally striking.

Unlike most parts of Latin America, Costa Rica was not settled by conquistadores establishing huge haciendas based on the labor of a suppressed Indian population. 3/ It also differs from most of the other contemporary less developed countries because most of its present population is descended from European settlers.

At the time of the Spanish conquest of Mexico and Central America, the region to become Costa Rica was sparsely populated. Moreover, there was no centralized Indian empire to subdue. Instead, the Spaniards were forced to contend with numerous small but fiercely independent groups of Indians. The climate and terrain also hindered exploration and settlement, but such efforts continued because of the promise of gold. In fact, no mineral wealth was discovered and Costa Rica remained the smallest and the poorest of the Central American colonies. Settlement was confined almost entirely to two highland valleys in the central plateau where the modern capital of San Jose is located. There were relatively few immigrants, and it was not until the nineteenth century that the population finally exceeded 20,000. Although some Spanish settlers took Indian wives, intermarriage among the Spaniards was the general rule and the population has remained largely European. Unlike other regions of Latin America, the Spanish settlers and their descendants in Costa Rica were obliged to do most of their own farm work and holdings were uniformly small.

Following independence from Spain in 1821, Costa Rica emphasized expanding production of coffee which has remained the principal commercial crop. In some recent years, export proceeds from bananas have exceeded receipts from coffee. Production of bananas for export is concentrated in foreign-owned plantations, although many independent growers have started production since the 1960s (Blutstein et al., 1970, p. 233). Until the mid-nineteenth century, Costa Rica's agriculture was dominated by small family farms. But, in the latter part of the century, there was a considerable concentration of land in relatively large coffee haciendas. 4/ Rising capital costs from more elaborate processing equipment encouraged this process. Many farmers purchased equipment on credit and then defaulted on their loans and lost their land. "The scarcity of labor," it is reported, "maintained a relatively equal distribution of income even in the face of growing inequalities in land ownership" (Saenz, 1972, p. 24). In brief, the demand for labor was sufficiently strong that competition for workers pushed up wages so that they received a share of the economic rent created by expansion of the coffee industry.

By the late nineteenth century, however, population growth began to exceed the rate of expansion of coffee production. A sharp drop in world coffee prices at the turn of the century and again in the 1930s depressed the demand for labor. Resulting declines in real wages were accentuated by relatively high food prices because little attention had been given to increasing food output. Protest demonstrations in the 1930s induced the government to permit occupation of public lands as a safety valve in a policy reminiscent of the opening up of the public domain to agricultural settlement in nineteenth-century United States. Much of the land available for colonization was located in the Pacific coast region and in the extensive Caribbean lowland region. By 1963, nearly half of the population was located in the more recently settled areas whereas as late as 1916 some 80 percent of the population was still concentrated in the central plateau. A change in the political orientation of the government dating from 1948 led to enlarged investment in infrastructure as well as expansion of social services and national-ization of the banking system (Saenz, 1972 pp. 32-39).

An agricultural development program initiated in 1970 gave high priority to increasing productivity and output among small-scale producers of maize, beans, and rice, principal commodities in the local diet. The agricultural "frontier" available for expanding culti-vated area was being rapidly reduced because of rapid population growth. Emphasis was, therefore, being given to intensification of production through high-yielding varieties, fertilizers, and improved agronomic practices (AID/San Jose, 1970, pp. 18-26).

Concentration of landownership in Costa Rica almost matches that of other countries of Latin America. The estimated 1960 Gini coefficient of 0.82 was only a little less than the coefficients of 0.83 in Guatemala and 0.87 in Colombia. According to estimates for 1963, fewer than 1,000 holdings of 500 manzanas or more accounted for some 42 percent of the total farm area but only about 1.5 percent of the total number of farms (Berry and Cline, 1979, pp. 38-39; AID/San Jose, 1970, p. 163). 5/

Despite concentration of land in large holdings, income distribution in Costa Rica appears to be relatively satisfactory. In 1961, the lower income 60 percent of the population received 24 percent of the

total income, compared with the 60-30 relationship in
Taiwan and 60-19 percent in Colombia that year.
Between 1961 and 1971, however, the per capita income
for households in the lower income 60 percent increased
at an impressive annual rate of 5.1 percent, whereas
the average rate of increase of per capita income for
the entire population was only 3.2 percent.
Consequently, the share of income accruing to the lower
income 60 percent in Costa Rica had risen to 28.4
percent by 1971 (Chenery, 1980, p. 29).

Rapid expansion of nonfarm employment in industry
and services was a major factor in improving the income
distribution situation in Costa Rica. The percentage
of the labor force in agriculture had already declined
to 51 percent in 1960. By 1979, agriculture's share in
the total labor force was about 30 percent. During the
same period, the percentage of the labor force in
industry rose from 19 to 23 percent and in services
from 30 to 47 percent. A substantial part of the labor
force in the service sector is in government employment
although separate figures are not available (Saenz,
1972, p. 56).

Malaysia

Considerable success has been achieved in Malaysia
in simultaneously implementing focus and dispersal
strategies. 6/ This success applies especially to
Peninsular (West) Malaysia; this brief account ignores
the experience of Sabah and Sarawak. Even though the
land area suitable for agricultural use in Sabah and
Sarawak is larger than in Peninsular Malaysia, total
population is less than one-fifth the population of
Peninsular Malaysia and economic development has been
much more limited.

Although coffee was the first major export crop
promoted during the British colonial period, the real
upsurge in agricultural development in West Malaysia
dates from the turn of the century when the demand for
rubber began to expand rapidly and rising prices
enhanced the profitability of this new crop. The
emphasis of the colonial administration was on the
expansion of plantation production. By 1921, area
planted to rubber in estates had already reached
765,000 acres. The Malaysian peasants, however, soon
responded to this new opportunity; by 1921, area
planted to rubber by smallholders had reached 470,000
acres or 38 percent of the total rubber area. Between
1921 and 1956, the year before Peninsular Malaysia
became independent, the total area planted to rubber
grew from 1.2 million to 3.5 million acres, and
smallholders accounted for 43 percent of the total
(Wafa, 1978, p. 38). This vigorous response of the

Malaysian smallholders is especially impressive because
the promotion efforts of the colonial regime "only
supported expatriate and particularly British
interests" (Barlow, 1978, p. 433). Not surprisingly,
government policy since independence has given a much
higher priority to the smallholder sector. By 1968,
smallholders accounted for just over 60 percent of the
4.3 million acres planted to rubber (Wafa, 1972, p.
43). However, smallholders accounted for a smaller
share of total production--54 percent--because of lower
yields (Barlow, 1978, p. 439).

The first European investments in Malaysia were in
tin mining. Rubber has, however, long been the
dominant economic activity in the country. As of 1970,
approximately 60 percent of the total cultivated area
was in rubber. Well over half of the agricultural labor
force was engaged in rubber production, although many
smallholders combined rubber production with other
agricultural activities. By 1970 the manufacturing
sector accounted for a slightly larger share of the
gross domestic product--13.1 percent compared with 12.9
percent for rubber--but it provided employment for only
293,000 workers compared with the 765,000 mainly dependent on rubber production. Rice, the overwhelmingly
important domestic food crop, accounts for roughly 20
percent of the cultivated area, followed closely by oil
palm. By 1973, acreage under oil palms represented
some 15 percent of the total area cultivated.

Two characteristics of Malaysia's agricultural
economy account for the dominant position of rubber and
oil palm: the rain forest climate and the plentiful
virgin lands.

Because of abundant rainfall and a continuously
warm tropical climate, virtually all of Peninsular
Malaysia is a rain forest zone. Exports of logs and
timber, expanding even more rapidly than exports of
palm oil and kernels, accounted for nearly 9 percent of
total export earnings in 1973, compared with 7 percent
for oil palm products. Rubber, accounting for 40
percent of total export revenue in 1973, was by far the
leading export, followed by tin with 15 percent of
foreign exchange earnings. Tree crops such as rubber
and oil palm are particularly well adapted to tropical
rain forest areas which satisfy their soil requirements. Annual crops and short-term perennials, unlike
tree crops, cannot hold a reasonable balance between
leaching and remobilization of nutrient elements.
Because of their extensive root system, tree crops are
able to make good use of a large part of the rainfall
throughout the whole year. And, they can store (in
their own structure or in their leaf litter) a
substantial portion of the nutrients available.

The other characteristic explaining the tremendous expansion of rubber and oil palm is the availability of plentiful and accessible virgin lands in the Malaysian Peninsula. The peninsula is still quite sparsely populated despite an enormous increase of population during the past century from natural increase and substantial immigration from China, India, and Indonesia. By 1900, light roads and railroads had been constructed to serve the tin mines and experience with the plantation or estate technique of organizing production had been acquired in the late nineteenth century with the establishment of a sugar industry. Both factors gave rubber and oil palm production a boost. In Malaysia, the term "estate" is commonly applied to units of 40 hectares or larger, although most of the rubber plantations are in fact much larger, currently averaging over 1,000 acres. They were, therefore, able to attract personnel with the needed managerial and technical expertise as well as capital. A "Loans to Planters" scheme was initiated in 1904, but overseas investors supplied most of the capital, financing the rapid growth of estate production of rubber. Profitability of rubber production and the political stability provided by a strong government, both in the colonial and postindependence periods, have favored this influx of foreign capital.

Since the late nineteenth century, research has made a critical contribution to the growth of the Malaysian rubber industry. Henry Ridley, a botanist with an exceptional ability to identify and solve practical problems, made notable contributions. When he became Director of the Botanical Gardens in Singapore in 1888, he focused his attention on the Hevea rubber trees planted earlier from stock imported from Brazil by way of the Kew Gardens and Ceylon. His pioneering work in developing techniques of tapping rubber trees and his enthusiastic efforts in distributing seeds and encouraging the new crop contributed notably to the establishment of the first plantations. A Department of Agriculture was established in 1905 with responsibility for rubber research. Some of the large commercial companies have also maintained research units. In 1926, the colonial administration established a Rubber Research Institute which took over government responsibility for research on rubber. The institute also had responsibility for extension activities and, in 1934, a "Smallholders Advisory Service" was created. The scope of research was expanded greatly after 1945, and the Rubber Research Institute became the largest research institute of its kind in the tropics (Barlow, 1978, p. 91). As of 1973 approximately half of its budget was devoted to research on production problems, including research on processing carried out in Malaysia. The Malaysian Rubber Producers' Research Association located in England has also done important work on processing and other problems important to promoting the use of natural rubber in industrial countries.

Rapid expansion of synthetic rubber production during World War II posed a new challenge for the Malaysian rubber industry. By 1946, synthetic rubber accounted for close to half of total world production of about 1.7 million tons. Production of synthetic rubber in 1950, however, had fallen to just over 500,000 tons and accounted for only 22 percent of world rubber production. Even that level of production depended on the United States enforcing a minimum level of synthetic rubber content in rubber goods because at that time the synthetics were not as useful as natural rubber.

Technological progress in the production and use of the various types of synthetic rubber, together with government measures to promote the synthetic industry in North America, Europe, and Japan, has since led to enormous growth of the industry. By 1973, production of synthetic rubber reached 7.3 million tons and represented 68 percent of world production. Malaysian and world production of natural rubber stagnated during the 1950s; world production of 2 million tons in 1960 was only marginally above the 1.9 million tons produced in 1950. But world production of natural rubber rose to 3.5 million tons in 1973. The increase in Malaysia was even more rapid, and its share in world production rose from 36 percent in 1960 to 43 percent in 1973 (Barlow, 1978, pp. 408, 444-45).

Viewing technological progress in the synthetic rubber industry, some observers in the 1950s and 1960s anticipated that synthetic rubber would be superior and cheaper and gradually displace the natural industry. The ability of Malaysia's rubber industry to survive and then achieve a rapid rate of growth after 1960 largely results from research on production and processing which reduced costs and improved quality. Achievements in plant breeding were especially significant, but those advances have been accompanied by noteworthy progress in agronomic practices, including use of fertilizers, and improvements in tapping practices. Advances in processing and packaging techniques have also improved the competitive position of natural rubber vis-a-vis synthetics. 7/ For example, problems related to the variable quality and characteristics of natural rubber have been overcome by schemes for producing technologically specified and conveniently packaged "block rubber." Moreover, the sharp rise in world petroleum prices since 1973 has tended to increase the cost of producing the major types of synthetic rubber and appears to have improved the competitive position of natural rubber.

The Malaysian government has generally not distorted the international prices passed on to Malaysian rubber producers. Export taxes and rubber cesses have been levied to mobilize support to the Rubber Research Institute and other development activities. Replanting schemes financed by the export mechanism enabled a large proportion of smallholders to replant old stands of rubber with higher yielding materials, a significant element in the government's dispersal strategy for rubber. (We note shortly that the important smallholder rubber sector in Indonesia has not benefited from this type of dispersal strategy.) Furthermore, the government has not resorted to the type of policies pursued by a number of export marketing boards in developing countries which have seriously dampened producer incentives.

An important weakness of the rubber research in Malaysia should be noted: research emphasis on estate production. Plant breeders, for example, have aimed at high yields under the low planting densities and high material input levels employed on estates with their relatively capital-intensive operations based on a hired labor force. In contrast, smallholders use high planting densities but much lower levels of fertilizer and other purchased inputs. An important example is that plant breeding to increase yields has been concentrated mainly on budgrafts and the evaluation of clones has been mainly carried out on estates; selection has therefore been biased toward those conditions. Some attention, however, has been given to selecting high-yield seedlings. Although their yield potential is not as great as for the most promising clones, selected seedlings have some advantages over budgrafts: a shorter period of immaturity, greater robustness, and higher initial yields (Barlow, 1978, p. 123). Those and related characteristics make use of seedlings more suitable for a dispersal strategy aimed at smallholders.

This choice between budgrafts and seedlings appears to be especially important to the design of rubber improvement programs in Indonesia. The major emphasis in the current rubber improvement schemes in the Outer Islands of Indonesia is on a focus approach even though the participants are small-scale farmers. This is because the improvement "package" is based on budgrafts and other sophisticated and costly components which means that only a relatively small number of farmers can participate. It would take well over 50 years to reach all of the small rubber producers at the current rate of expansion. And a substantial acceleration of that rate with the focus strategy would be impossible because the demands for trained manpower and financial assistance are so large (Barlow and Johnston, draft manuscript). Budgrafts may not survive without good management, and they perform well only with high levels of fertilizer application and weed control. Being more robust, selected seedlings can withstand relatively poor management, although they respond well to better conditions. Thus they lend themselves to a dispersal strategy involving progressive adoption of improved practices as knowledge is diffused and skills are upgraded. In fact, in the cases of Malaysia and Thailand, many smallholders have already planted budgrafts.

The bias in research against smallholders has been reduced in Malaysia since independence. Furthermore, a number of positive measures have been adopted to expand employment and income-earning opportunities for the rural population. One of the most ambitious initiatives was the creation in 1956 of the Federal Land Development Authority commonly known as Felda. This was a generously financed scheme to promote more rapid opening up and settlement of land in underdeveloped parts of the country, mainly for rubber or oil palm production. The Felda schemes have been criticized as being overly expensive, too centralized, and too tightly controlled by a hierarchy of managers, assistants, and supervisors (Barlow, 1978, p. 235). A number of settlement schemes have also been initiated by state governments; these often have had advantages because they are more "settler-oriented" (Wafa, 1972, pp. 256-63).

Measures to achieve rapid expansion of rice production have been another highly significant dispersal program. In 1956, on the eve of independence, Malaysia imported 45 percent of its rice. The colonial administration emphasized policies to promote cheap labor and cheap food. Some efforts were made to help farmers improve rice production in order to increase self-sufficency, but in fact there was heavy reliance on imports from Thailand and Burma.

Following independence, an ambitious drive to achieve self-sufficiency in rice was launched. By 1971, imports represented only about 13 percent of total consumption. Between 1956 and 1970, $161 million--36 percent of total expenditure on agricultural development--was spent on irrigation and drainage projects. Virtually all of this was for rice and it permitted a large increase in double cropping. During the decade ending in 1971-72, Malaysia's rice production increased by 60 percent, made possible largely by a tremendous expansion in the off-season crop which accounted for only 5 percent of total production in 1962 but 40 percent in the 1971-72 crop year.

In addition to large, well-managed investments in the construction of irrigation and drainage systems, rapid adoption of high-yield, quick-maturing varieties made a notable contribution to the expansion of rice production. IR-5 and a cross between a local variety and a Taiwanese variety, Taichung 65, were among the first to spread rapidly. The quick-maturity feature of the new varieties was critical to the expansion of double cropping. The increase in multiple cropping and expansion of output in turn resulted in a substantial increase in demand for labor. Tightening of agricultural labor markets was especially apparent in areas served by the Muda Scheme, largest of the irrigation and drainage projects. Harvest wage rates in this area increased by 80 percent between 1971 and 1974, according to rough calculations by Goldman (1975, p. 268). This was considerably greater than the 50-percent increase in land rents, striking in that the direct effects of new varieties and improved irrigation were to sharply enhance the productivity and value of land.

A substantial subsidy on fertilizers and government policies affecting rice prices have also influenced expansion of rice production. Farm prices have been supported mainly by restricting imports, but government purchases for building stocks are made when the price of certain grades of paddy rice fall below the support level. In the absence of protection, the consumer price of rice would have been nearly 20 percent lower than its observed level, according to Goldman (1975, p. 286). Therefore, beneficial effects of the rice policy on incomes of many low-income farm households was offset to some extent by the implicit tax borne by families depending on purchased rice.

There is a cogent equity case for the post-independence rice policies which had aimed at improving incomes and employment opportunities for rice farmers. Household surveys carried out in the late 1960s indicated that single-crop rice farmers were earning a net income (including home consumption) of only about $100 from rice and a total income of some $200 to $230, compared with an income of close to $1,300 for the average Malaysian family. The programs to expand rice production and to increase double cropping led to a significant expansion in the demand for labor and an increase in wage rates which significantly improved incomes of many of the poorest rural households.

In the case of the large Muda Project, the increase in farm income between 1967 and 1974 was striking. According to rough estimates by type of household, real per capita incomes of the households with farms in the command area increased by approximately 85 percent during that period. And, the landless households engaged in rice production increased their per capita incomes by the same percentage although the increase was from a considerably lower base. The increase in

income among farm households outside the project area
and among nonfarm households was less but still a
substantial 50-percent increase (Bell and Hazell, 1980,
p. 82). Final demand linkages appear to have been much
more important than backward or forward linkages in
accounting for "downstream effects" of the project on
nonfarm incomes in the region. In 1974, per capita
incomes of the nonfarm population in the region were
two to four times the level of incomes among farm
households and landless families even though the
percentage increase they registered between 1967 and
1974 was only about 60 percent as great as for the rice
farmers in the project.

Writing at the end of the 1967-74 period when the
Muda Project reached its initial goals, Goldman (1975,
p. 289) expressed concern that mechanization of rice
harvesting would have adverse effects on the demand for
labor and labor incomes: "If further mechanization
takes place, therefore, its likely impact will be to
capitalize into land values the rents heretofore
captured by labor. This result would be ironic indeed
for a rice policy whose major goal is augmenting
incomes of the rural poor."

A recent analysis of the Muda Region by Bell,
Hazell, and Slade, reports that small Japanese combine
harvesters were already harvesting some 35 percent of
the main season crop in 1977/78. By means of a semi-
input-output model, the authors attempted to quantify
the effects on employment of the use of combines to
harvest 60 percent of the total area (the maximum
considered feasible for technical reasons related to
topography and irrigation system). According to
estimates derived from their model, wage rates with
mechanized harvesting would be about $3 per day
compared to well over $4 a day in the absence of
combines. The most significant difference, however, is
that the amount of daily labor hired would be reduced
so sharply that total wage payments to landless workers
and seasonal migrants would fall to about $2 million
compared with some $15 million in 1974 (Bell, Hazell,
and Slade, forthcoming, pp. 11-13).

Landowning households are, of course, investing in
combines to minimize production costs. There appear to
be three principal reasons why public policy has not
attempted to curb investments in this highly labor-
displacing innovation. First, the number of landless
households affected is relatively small, a little over
3 percent of the 125,000 households in the Muda Region
in 1972 (Bell and Hazell, 1980, p. 77). Second, many
of the day laborers losing employment oportuntities are
migrant workers from Thailand or Indonesia with little
political influence. Third, because of the considerable growth of employment opportunities in Malaysia's

manufacturing sector, "problems of labor scarcity have become increasingly severe in both the peak cultivation and harvesting periods" (Barlow, personal communication, Dec. 22, 1981). Labor shortages are also emerging in the smallholder rubber sector. Barlow reports that large numbers of "illegal" emigrants from Indonesia are working on construction sites, in the estate sector, and to a lesser extent in the smallholder sectors.

Malaysia is unique in the extent to which large increases in output in the estate sector have been paralleled by significant increases in income among a large and growing population of smallholders. This continuing DSS structure is characterized by large inequalities in income. The data in table 3.1 indicate that the share of income accruing to the highest 20 percent of households was some 17 times as large as the share received by the bottom quintile.

These rough estimates should be treated cautiously. There seems little doubt, however, that income distribution in Malaysia is considerably less equal than in Taiwan and Korea. On the other hand, the situation in Malaysia is similar to Costa Rica and the share accruing to the bottom 20 percent was appreciably higher than in Mexico. Although the estimated share of total income received by the bottom 20 percent in India is substantially greater than in Malaysia, the absolute level of income received by the lowest quintile of households in Malaysia is much higher. Average per capita income in Malaysia was six times as high as in India whereas the differential in the share of income accruing to the bottom 20 percent was 2 to 1.

The DSS pattern of rural development in Malaysia has certainly not been optimal from an equity point of view. The parallel emphasis on focus and dispersal strategies has, however, had important advantages as well as disadvantages. Because the estate sector has been highly successful in attracting capital from overseas investors, the overall rate of growth of output has undoubtedly been more rapid than would have been possible with a development strategy restricted to dispersal strategies favoring smallholders. Moreover, there have been other benefits including investments in infrastructure and in the accumulation of technical knowledge by research organizations, notably the Rubber Research Institute, and from experience in the estate sector that facilitated the expansion of output of rubber and oil palm among smallholders. And the more rapid growth of tax revenues has helped to finance expansion of social services.

However, the fact that the focus and dispersal strategies were to a large extent complementary rather than competitive was a result of a special combination

TABLE 3.1

Percentage share of household income by quintile groups of households

Country and Year of Estimate	Average per Capita Income in 1978 U.S. dollars	Lowest 20 percent %	2nd Qunitile %	3rd Quintile %	4th Quintile %	Highest 20 percent %
Malaysia (1970)	1,090	3.3	7.3	12.2	20.7	56.6
Costa Rica (1971)	1,540	3.3	8.7	13.3	19.9	54.8
Mexico (1977)	1,290	2.9	7.0	12.0	20.4	57.7
Taiwan (1971)	1,400	8.7	13.2	16.6	22.3	39.2
Rep. of Korea (1976)	1,160	5.7	11.2	15.4	22.4	45.3
India (1964–66)	180	6.7	10.5	14.3	19.6	48.9

Source: World Bank, 1980, pp. 156–57.

of circumstances. Most obvious is the fact that the
political power of the Malay community has mitigated
the biases against the smallholder sector. Even more
important are certain features of Malaysia's resource
endowment and characteristics of the major crops which
have prevented the rapid expansion of production in the
large-scale sector from compromising the possibilities
for parallel expansion of the smallholder sector. We
examine those distinctive characteristics of Malaysia's
experience in the following section where we identify
the factors most useful in defining typologies of
development situations.

Various measures of education, health, and other
social services confirm that rural poverty has been
substantially reduced in Malaysia. Results in Malaysia
are not as impressive as in Costa Rica. But, Malaysia
has been independent only since 1957 and 50 percent of
the labor force still depended on agriculture in 1978,
compared with only 29 percent in Costa Rica. Furthermore, since the averages include Sabah and Sarawak, the
situation in Peninsular Malaysia is no doubt better
than suggested by national averages. In 1977, an
estimated 93 percent of the primary school age
population was enrolled in school. The percentage of
the secondary school age population enrolled increased
sharply from 19 percent in 1960 to 43 percent in 1977.
Only about 3 percent of the population aged 20-24 was
enrolled in higher education in 1976, but that was a
threefold increase since 1960. In spite of this
expansion of schooling, however, the adult literacy
rate in 1975 was estimated at only 60 percent, still
much above India's estimated rate of 36 percent.

Health programs in Malaysia have benefited rural as
well as urban people. The crude death rate, estimated
at 9 per 1,000 in 1960, declined to 6 per 1,000 in
1978. The estimated infant mortality rate in 1978 was
at the relatively low level of 28 per 1,000. The death
rate among children aged 1 to 4 was reduced from 9 per
1,000 in 1960 to 3 per 1,000 in 1970. These improvements were associated with a decline in the crude
birthrate from 39 to 29 per 1,000 between 1960 and
1978. Because this decline was considerably larger
than the decline in the crude death rate, the annual
rate of natural increase was reduced from 3 to 2.3
percent.

Diversity of Problems and Constraints and the Need for Typologies of Developing Countries

Countries examined in the preceding section
indicate the diversity of conditions influencing development problems and possibilities which should be
considered in the design and implementation of

agricultural strategies. Specific characteristics of each country and its agricultural regions are central to planning. But, the task of deriving useful lessons from experience can be facilitated by defining typologies of countries with similar characteristics.

Numerous parameters can be used to define country typologies: per capita income; the share of agriculture in a country's total labor force; resource endowment; size as measured by area, population, or GNP; historical experience and cultural factors; political regime or economic ideology; and the nature of a country's economic policies (e.g., inward- vs. outward-looking economic policies). These and many other factors can be important. But, multiplying the number of parameters leads to an exponential increase in the number of typologies to be considered and thus defeats the purpose of the exercise.

Given our emphasis on the choice between DSS and USF models of agricultural development, a useful start is to examine the four factors which made it possible for Malaysia to simultaneously pursue focus and dispersal strategies. A strong emphasis on focus strategies leading to a DSS pattern of development will preclude the possibility of implementing dispersal strategies and achieving an USF pattern of development characterized by widespread increases in farm productivity, employment, and income. The opportunity cost of encouraging the DSS model in terms of pre-empting the possibility of achieving a USF pattern of development depends primarily on the following factors: (1) land availability; (2) the cash income or purchasing power constraint; (3) the economic and technical characteristics of the country's major crops; and (4) the scarcity of capital, foreign exchange, government funds, and trained manpower. The uniqueness of conditions in Malaysia is apparent in relation to all of those factors.

(1) Land availability. Abundance of virgin land in Malaysia has meant that the large land areas of the estate sector have not significantly restricted expansion of the smallholder area. Conversely, when agricultural area is small relative to the number of farm households, as in India, Bangladesh, or Kenya, expansion of large-scale enterprises inevitably reduces the land available for small-scale farm units. Hence, the great majority of farms will be exceedingly small. As population pressure increases, a growing percentage of the rural population will become virtually landless.

(2) The cash income or purchasing power constraint. The estate sector in Malaysia is concentrated on the production of rubber and oil palm for export; therefore, the large-scale sector does not interfere with efforts to increase farm cash receipts

of the small-scale sector. But, increasing cash receipts for the average farm unit will be generally constrained by the small size of the domestic commercial market. Some 50 to 80 percent of the population still depends on agriculture and a relatively small fraction of the population relies on purchased food. Because of that cash income constraint, the purchasing power of farm households for acquiring fertilizers and other off-farm inputs is severely limited. That purchasing power constraint is accentuated if the large-scale subsector satisfies most of the commercial demand growth as industrial and urban growth modify the predominantly agrarian character of the economy. In the case of Malaysia, this cash income constraint has not been significant for two reasons. First, to a large extent, the growth of output in the estate and smallholder sectors has been oriented toward export production. Second, the possibility for rapidly expanding the cash receipts of rice farmers has been large as Malaysia implemented development programs and protectionist policies promoting rice self-sufficiency.

(3) <u>Economic and technical characteristics of the country's major crops</u>. When a developing country accounts for a substantial fraction of world exports of a commodity, a sizable increase in export production is likely to have an adverse effect on export prices because export demand for many export crops is so inelastic. Ghana's experience in rapidly expanding cocoa exports during the early 1960s is an important example. The demand schedule facing Ghana's cocoa exports was approximately of unit elasticity so that the large increase in volume was offset by the decline in cocoa export prices. The expansion of synthetic rubber production since World War II reduced demand for natural rubber production. But such production substantially increased the price elasticity of export demand facing Malaysia and other exporting countries. In the case of palm oil, the extremely rapid expansion of Malaysia's exports—virtually a fivefold increase between 1969-71 and 1979—had a relatively small impact on export prices because of the great importance of substitution among vegetable oils. An important technical characteristic of rubber, oil palm, coffee, and other tree crops is also worth mentioning: because of their dependence on hired labor, estates use less labor-intensive technologies than smallholders. However, the scope for labor-saving mechanization is much less for those three crops than for field crops like wheat or soybeans. Consequently, a given rate of expansion of production can be expected to generate a considerably larger increase in demand for labor in a tree crop like rubber or coffee than in a crop like wheat which lends itself to mechanized production.

(4) <u>Scarcity of capital, foreign exchange, government funds, and trained manpower</u>. In any country, there are competitive tradeoffs in the allocation of scarce resources of capital, foreign exchange, government funds, and trained manpower between focus strategies and dispersal strategies. In the case of Malaysia, however, the estate sector was able to obtain most of its capital from overseas investors so that the funds were largely a net addition to development funding. During the earlier stages of development of Malaysia's rubber and oil palm industries, much of the managerial and technical expertise was also recruited abroad. In later periods, some of the manpower trained in the estate sector probably became available to augment the administrative capacity of the government for implementing dispersal strategies. Foreign exchange earnings and tax payments of the rubber and oil palm estates eased the country's foreign exchange and budgetary constraints.

Major Factors Differentiating Developing Countries

The foregoing discussion of the special circumstances enabling Malaysia to simultaneously implement focus and dispersal strategies directs attention to three factors: (1) per capita income as an indicator of the severity of financial and administrative constraints; (2) the degree of structural transformation, especially the share of agriculture in the country's total labor force; and (3) the nature of a country's resource endowment. The significance of the first two factors is enhanced because, with few exceptions, low-income developing countries are also characterized by the large share of their labor force still dependent on agriculture. The nature of a country's resource endowment obviously influences the level of per capita income. This is most obvious in the case of the so-called "capital-surplus oil exporters." It is, however, the availability of agricultural land for expanding the cultivated area that is particularly relevant to the choice between the DSS and USF models. In addition, the resource endowment also influences the potential for increasing farm employment opportunities and incomes by expanding rapidly the production of export crops.

In its World Development Report, 1980, the World Bank gives estimates of a variety of socioeconomic statistics for 38 low-income developing countries and 52 middle-income developing countries. 8/ Table 3.2 summarizes the basic indicators for those two categories plus the group averages for 18 high-income, industrialized countries; data for three low-income and three middle-income countries discussed earlier are

also included. The middle-income developing countries
are a heterogeneous group but the three countries we
have considered--Taiwan, Costa Rica, and Malaysia--all
fall in the middle range in terms of per capita income
for that group of countries. The more than sixfold
differential between the average per capita income of
the middle-income countries and the low-income countries in 1978 has been influenced considerably by their
more rapid growth of per capita GNP between 1960 and
1978: an average rate of 3.7 percent compared to only
1.6 percent as an average for the low-income countries.
     The differential between the low- and middle-income
developing countries in the share of agriculture in the
total labor force was large in 1960 and further widened
between 1960 and 1978. Many factors influence the
occupational composition of a country's labor force and
also the extent to which the population of working age
participates in economic activity (Standing, 1978). The
change in the absolute and relative size of a country's
farm labor force is, however, influenced powerfully by
the initial weight of agriculture in the total labor
force and by the rate of increase in the total labor
force. 9/
     The significance of the distinctive characteristics
of the subset of low-income, late-developing countries
is accentuated because of the persistence of high rates
of natural increase in most of these countries. Sri
Lanka, with a crude birth rate of only 26 per 1,000 in
1978, is the only low-income developing country that
has clearly entered the stage of the demographic
transition characterized by a substantial decline in
fertility. Two of the most populous low-income
countries--India and Indonesia--also registered a
considerable decline in fertility between 1960 and
1978. Reductions there account for much of the decline
in the population-weighted average crude birthrate in
low-income countries shown in table 3.3 The fact that
the decline in the unweighted average crude birthrate
was only from 47 to 45 per 1,000 emphasizes the fact
that most of the low-income countries have experienced
virtually no reduction in fertility. There is a lag of
some 15 years before a decline in fertility is
reflected in a reduction in the rate of growth of a
country's labor force. Therefore, declines in the
share of agriculture in the total labor force will
continue to be slow because of the rapid growth of
their population of working age in combination with
the heavy weight of agriculture in their total labor
force.
     Many middle-income countries with relatively low
per capita GNP continue to have high rates of fertility
and also a large share of their labor force in agriculture. Of 28 countries with 1978 per capita GNP less

TABLE 3.2
Low-income and middle-income developing countries: Basic indicators

Category and Country	Population, 1978 Number	Per Capita GNP, 1978 U.S. dollars	Average Annual Growth Rate of GNP %	Percentage of Labor Force in Agriculture 1960 %	1978 %
38 low-income developing countries:					
Group average--	1.2 billion (total)	200	1.6	77	72
India	643.9 million	180	1.4	74	74
Kenya	14.7 million	330	2.2	86	79
Tanzania	16.9 million	230	2.7	89	83
52 middle-income developing countries:					
Group average--	873 million (total)	1,250	3.7	58	45
Taiwan	17.1 million	1,400	6.6	56	37
Costa Rica	2.1 million	1,540	3.3	51	29
Malaysia	13.3 million	1,090	3.9	63	50
18 industrialized countries:					
Group average--	668 million	8,070	3.7	17	6

Source: World Bank (1980).

Note: The range of GNP per capita is from $90 to $360 for the low-income countries and from $390 to $3,500 for the middle-income group.

TABLE 3.3
Fertility, mortality, and rates of natural increase in population
in low- and middle-income developing countries

Category and Country	Crude Birth Rate 1960	Crude Birth Rate 1978	Crude Death Rate 1960	Crude Death Rate 1978	Rate of Natural Increase 1960	Rate of Natural Increase 1978
	Number per 1,000				%	%
Low-income developing countries:						
Group average--						
Unweighted	47	45				
Population-weighted	48	39	24	15	2.4	2.4
India	43	35	21	14	2.2	2.1
Kenya	51	51	19	14	3.2	3.7
Tanzania	48	48	22	16	2.6	3.2
Middle-income developing countries:						
Group average--						
Population weighted	40	35	14	11	2.6	2.4
Taiwan	39	21	7	5	3.2	1.6
Costa Rica	47	28	10	5	3.7	2.3
Malaysia	39	29	9	6	3.0	2.3

Source: World Bank (1980).

than Malaysia's per capita GNP of $1,090 only 4 had a
crude birth rate in 1975 of 35 per 1,000 or less
(Thailand, the Philippines, Colombia, and Tunisia).
And, in only 8 of the 28 had the share of agriculture
in the total labor force in 1978 declined to 50 percent
or less. In contrast, virtually all the middle-income
countries ranking above Malaysia's per capita GNP
registered substantial declines in fertility by 1978.
The decline in agriculture's share in the total labor
force in this more prosperous group of middle-income
countries was also substantial. Only Turkey with 60
percent of its labor force in agriculture had a higher
share than Malaysia in 1978.

Implications of Economic, Structural, and Demographic Constraints for Agricultural Development
Strategies. The "low" and "lower middle" income
developing countries face severe resource constraints
limiting feasible development options. Moreover, the
existing economic and occupational structure in these
countries emphasizes the predominantly rural character
of their problems of poverty. The persistence of high
rates of fertility in combination with declining
mortality rates means that rapid population growth will
continue to compound the difficulty of reducing poverty
in these countries and of transforming the predominantly rural character of their economies.

Success in accelerating the growth of agricultural
production in these low-income, late-developing
countries is a necessary but not sufficient condition
for reducing malnutrition and other serious and
widespread manifestations of poverty. In order for
development to lead to widespread increases in
productivity and income, there is an urgent need to
design and implement effective dispersal strategies
leading to a USF pattern of agricultural development.
Unless a country is favored by the very special
combination of circumstances that shaped Malaysia's
development possibilities, emphasis on focus strategies
and a DSS pattern of agricultural development will
almost certainly mean that poverty among a large
fraction of their population will persist and will be
accentuated as growing population pressures hinder
efforts to raise per capita farm incomes.

The enormous importance of agricultural strategies
that permit widespread increases in productivity,
employment, and income among farm households is further
emphasized by the limited scope for alleviating poverty
by relief and welfare measures. That is, the severity
of the constraints facing these countries rules out the
large resource transfers required to reduce such
pervasive poverty. Some middle-income countries, for
example, Mexico or Brazil, have sufficient resources to
mount redistributive schemes such as food subsidy

programs to raise directly food consumption levels of poor families. These are semi-industrialized, middle-income countries where because of a rapid reduction in fertility and the rate of growth of population and labor force the demographic situation is changing. Because of increased per capita GNP and structural transformation of their economies, the purchasing power constraint, as well as financial and manpower constraint (typical of developing countries) have been reduced. Even in these countries, design and implementation of a dispersal strategy should receive a high priority in certain regions. These are specific areas where poor, small farmers are concentrated; regions of this kind have often been bypassed in the development progress of these economies, as witness, for instance, the High Tell of central Tunisia. But, for the low-income, predominantly agricultural countries, such schemes offer little hope. Because of severe resource constraints in such countries, the opportunity cost of diverting resources of money and manpower from the challenging task of promoting widespread agricultural development is bound to be high.

Nonagricultural Components of Agricultural and Rural Development Strategies

This book emphasizes production programs aimed at fostering widespread increases in agricultural productivity, employment, and incomes. It is essential also to recognize some highly significant relationships between those production-oriented programs and other elements of a rural development strategy. These include (1) the influence of the pattern of agricultural development on education and human capital formation; (2) the growth of rural-based industry and other measures to expand nonfarm employment opportunities; and (3) social service programs aimed at strengthening rural schooling and rural health and family planning programs.

There are inevitably tradeoffs in the competition for scarce resources between agricultural programs and rural education and health programs. But, there are also significant complementarities. We examine those interrelationships in chapter V when we confront the problem of determining priorities in the allocation of scarce resources. At this point, we introduce several important complementarities among the elements of a country's agricultural and rural development strategies. First, the experience of Japan and Taiwan suggests that a USF pattern of agricultural development tends to create an environment favorable to the expansion and strengthening of rural schools and also to human capital formation based on widespread opportunities for "learning by doing." In addition, the widespread involvement of the farm population in technical and economic change under a USF pattern appears to create a more propitious environment for the spread of family planning than a DSS pattern in which the majority of the farm population is bypassed.

Second, the pattern of rural demand associated with the USF model stimulates more rapid growth of output and especially of employment in nonfarm industries than with a DSS pattern. That is, more widespread growth of farm incomes and purchasing power generates demand for relatively simple consumer goods and farm implements. Such products tend to be produced by small- and medium-scale manufacturing firms. These firms employ more labor-intensive technologies than those employed by the larger and more capital-intensive firms producing consumer goods and farm machinery for the large-scale enterprises which account for the bulk of commercial sales in a DSS pattern. In addition, many of the more sophisticated items purchased by wealthier farm households--tractors or automobiles, for example-- are often imported. Even when such items are produced locally, little is involved beyond the assembly of imported components. Domestic value added and employment generated are limited.

We examine the evidence related to those linkages in more detail in chapter IV. Although agriculture must absorb a large fraction of the annual increments to the labor force in low-income, late-developing countries, the expansion of nonfarm employment opportunities must play an increasingly important role in expanding the demand for labor so as to raise its opportunity cost and push up returns to labor. Generating additional employment opportunities through rural works programs has an important role to play in alleviating poverty, especially where the increase in the demand for labor fails to pace the growth of the labor force. Therefore, investment programs to improve the rural infrastructure considered in chapter IV are important as a source of job opportunities as well as a means to increase agricultural production.

Third, there are important complementarities between production programs and social service programs related to education and health. Investments in education have highly significant effects on human capital formation and on economic growth as well as on the well-being of those receiving formal education. Expanding educational opportunities, especially for women, also makes an important contribution to family planning.

Rural health and family planning programs also represent important "public goods." These programs can be provided more efficiently and with greater benefits to society when handled as social service programs rather than simply in response to the growth of private demand for health and family planning services. This is particularly true of low-cost health programs which emphasize preventive activities such as childhood immunization and nutrition education supporting breast feeding and better child feeding practices.

Despite substantial reduction in overall mortality, high mortality rates, chronic illness, and malnutrition among infants and small children still characterize many less-developed countries, especially the low-income developing countries. Well-designed rural health programs focused on infants and small children and their mothers can achieve substantial and rapid reductions in mortality and morbidity among those vulnerable groups at an affordable cost. Improved physical and cognitive development during the critical birth to age 5 period also has significant effects on the later performance of children 10/ The most fundamental factor influencing nutritional status and health is the availability of food as determined by home production or purchasing power--hence, our emphasis throughout this paper on the rate and pattern of agricultural development. In the case of infants and small children, however, it is inefficient to focus only on factors influencing the availability of food because of the great importance of the two-way interactions between nutritional status and health. Thus, a health intervention such as promoting the simple technique of oral rehydration often merits a high priority as an extremely cost-effective means of improving the nutritional status and health of this vulnerable category.

Improving survival prospects of children--and increased awareness of the improved prospects for child survival--is critical to efforts to reduce fertility and slow the rapid population growth The most dramatic benefits of slowing population growth are realized in the long run. But even in the short run, reducing average family size can raise food consumption levels and reduce malnutrition as well as reduce the resource requirements for expanding coverage of education and health programs. Furthermore, achieving rapid progress in lowering fertility levels is critically important for slowing labor force growth. Continued fast growth of the labor force will make it increasingly difficult to tighten labor supply/demand conditions in efforts to increase returns to labor. Without reasonable success in family planning, the relative surplus of unskilled labor will persist and even increase and the depressing effect of this surplus on labor incomes will be exacerbated. Disturbing is the fact that fertility has declined mainly among the urban and wealthier population; experience in Colombia is an example (Potter et al., 1976).

Rapid growth of the total labor force in the predominantly agricultural countries will inevitably mean large increases in the farm population and labor force. This will mean further subdivision of holdings and increases in the number of landless households. This situation will entail not only "diminishing returns" in agriculture, but also destructive use of croplands, forests, and watersheds as illustrated by the problems emerging in Kenya's semiarid farming areas.

Identification of Functional Areas for Analyzing the Potential for Generating Employment Opportunities for the Rural Poor

The rate and pattern of agricultural development depends upon a large number of interacting factors. It has become fashionable to blame the political power structure and lack of political "will" as decisive factors accounting for the failure to design and implement broadly based, employment-oriented agricultural strategies (Griffin, 1979). Inasmuch as alternative strategies have very different effects on "who gets what, when, and how," political factors are obviously important. But political factors exert their influence in specific functional areas. We believe, moreover, that the shortcomings of development efforts must be attributed as much to faulty analysis and to organizational and other weaknesses in implementation as to the nature of the power structure. The success of the USF patterns in Japan and Taiwan is hardly explained in terms of superior sensitivity on the part of their political leaders to the plight of the rural poor. Conversely, the impressive commitment on the part of Tanzania's President Nyerere and other leaders to advancing the well-being of the rural poor has not yet been translated into effective policies and programs.

We believe that prospects for success in design and implementation of a broad-based USF pattern will be determined to a large extent by decisions and investments related to seven functional areas: (1) asset distribution and access; (2) planning and policy analysis; (3) development and diffusion of new technology; (4) investments in rural infrastructure; (5) policies and programs related to marketing and storage, input supply and credit; (6) rural industry and ancillary activities; and (7) institutional development (improving organizational structures and managerial procedures).

We next analyze these functional areas to assess the potential impact of policies and programs on employment generation and the growth of incomes and purchasing power in countries with USF and DSS patterns.

NOTES

1/ This dichotomy between focus and dispersal strategies was introduced by Colin Barlow (1979) in the context of assessing alternative strategies for the development of rubber and other tree crops in Malaysia and Indonesia.

2/ The Arusha Declaration and the essay are included in Nyerere (1968).

3/ To some extent, this was also true of several other Latin American countries, perhaps most notably Paraguay and Nicaragua.

4/ A contemporary observer reported in 1851 that two-thirds of the population was made up of landowners (Saenz, 1972, p. 15).

5/ One manzana is equal to 1.72 acres.

6/ Malaysia is also attractive as a case study because of the rich literature documenting its experience. A major book by Barlow (1978) on the Malaysian rubber industry is especially valuable, in part because of its historical perspective and its attention to alternative crops as well as rubber. A doctoral dissertation on land development strategies in Malaysia by the late Syed Hussain Wafa (1972) is a valuable supplement to Barlow's treatment of settlement schemes. This account also draws heavily on an excellent article by Goldman (1975) on Malaysian rice policy with an emphasis on its impact on employment and income distribution. A Ph.D. thesis by Mokhtar Tamin (1978) has also been valuable for our treatment of the government's dispersal strategy for rice.

7/ The enormous expansion of oil palm production in recent years has also been based on research that developed high-yield varieties. Most of the research, however, was carried out in the former Belgian Congo and the Ivory Coast.

8/ The report for 1981 uses a classification differing from the 1980 report and does not include data for Taiwan.

9/ This is emphasized by the well-known identity which gives the rate of change in the farm labor force ($L'a$) as a "dependent variable" determined "exogenously" by agriculture's initial share in the labor force (La/Lt) and the rates of change in the total labor force ($L't$) and in the non-agricultural labor force ($L'n$):

$$L'a = (L't L'n) \frac{1}{La/Lt} + L'n$$

10/ The health and nutritional status of adults may have an important influence on the duration and intensity of their work (see especially Standing, 1978, chapter 4). The evidence is persuasive, however, that malnutrition and excessive morbidity among infants and small children is a more serious and widespread problem and one that can be alleviated quite dramatically by very cost-effective intervention programs. Programs for the control of malaria and tuberculosis appear to be the only interventions that are similarly cost-effective in improving the health of adults.

IV
Analysis of "Functional Areas" for Their Potential Impact on Employment Generation and Growth of Effective Demand

Chapters II and III examined the agricultural development experience of countries characterized by the DSS and USF models and the experience of four countries exhibiting different types of a "mixed characteristics model." Two major conclusions are suggested. First, broadly-based agricultural development is much more effective in expanding employment opportunities and generating income and effective demand than a dualistic pattern of development. Second, unless there are very special circumstances, as in Malaysia, the USF and DSS patterns of agricultural development tend to be mutually exclusive. That is, an emphasis on the "focus strategies" which increase the likelihood that a country's pattern of agricultural development will be dualistic tends to preclude the possibility of designing and implementing "dispersal strategies." A successful USF pattern of agricultural development is impossible without such dispersal strategies.

The fact that so few countries have successfully designed and implemented a USF pattern underscores the difficulty of the task. Political opposition to USF policies can be harmful, but cannot be blamed exclusively for failures of the development process. Shortcomings of policies and programs are also related to analytical and administrative deficiencies. Given the complexity and challenges of development problems, serious shortcomings in the design and implementation of agricultural development are hardly surprising. The challenge confronting policymakers in developing countries and assistance agencies is to learn from the mistakes and shortcomings revealed by experience. There is now much greater awareness of both the desirability and the feasibility of the USF model of agricultural development with its emphasis on labor-using, capital-saving technologies. Progress in translating that awareness into policies and programs effective in spurring agricultural production within a USF pattern is, however, limited.

This chapter examines policies and programs in the seven functional areas identified in chapter III. Our emphasis is on the potential impact of policies, programs, and investments in these functional areas on the expansion of employment opportunities and the growth of effective demand. The various functional areas are, of course, interrelated, but we defer our examination of those interrelationships until chapter V where we confront the problem of determining priorities.

An essential feature of a successful USF strategy concerns the interaction between farm-level factors and socially determined factors. The rate of growth of farm productivity and output depends on the decisions and performance of individual farmers; the skill, energy, and efficiency with which they combine land, labor, knowledge, and a variety of purchased inputs are the proximate determinants of the level of farm output. The ability and willingness of individual farmers to expand agricultural production will, however, be determined in large measure by the nature of the technological innovations that are generated by agricultural research programs, extension programs, and rural education. Such programs influence the diffusion of knowledge and the ability of farmers to make output-maximizing, cost-minimizing decisions. Government policies affecting farmers' access to resources and production incentives also heavily influence farm output.

A great institutional innovation of the nineteenth century was establishment of publicly-supported programs of agricultural research and extension which provide "public goods" such as high-yield, fertilizer-responsive crop varieties and a better understanding of agronomic practices. Even large farms are much too small to support the agricultural experiment stations. Moreover, inventions such as a new, high-yield variety of rice or wheat quickly become part of the public domain. Private firms therefore cannot capture the profits of innovation through patent rights. Consequently, private investment in most of the research and development activities critical to agricultural progress would be much below the level of investment that is socially desirable.

The first publicly-supported agricultural research station was established in Saxony in 1852. But, it was the United States and Japan that pioneered this new approach to providing the publicly-supported R&D activities (Hayami and Ruttan, 1971, chapter 7). Progress in strengthening rural public schools has also provided a "public good" that justified a considerably higher level of investment in education than would have been made if education had been provided only by private institutions. The skills of "reading, writing, and arithmetic" were important; but equally important were the effects of education in increasing the receptivity of future farmers to innovations and in enhancing their capacity to make farm management decisions. For many of the young people reared in farm households, this increased competence and ability to learn and innovate enhance their performance in a variety of nonfarm occupations as they seek employment outside the agricultural sector.

Our analysis of the impact of investments in the seven functional areas on the expansion of employment and of effective demand focuses on low-income, late-developing countries where rural poverty is most pervasive and intractable. For middle-income, semi-industrialized countries such as Brazil and Mexico, the expanding opportunities for employment in manufacturing and other nonfarm sectors can make a much larger contribution to the growth of employment opportunities. Because resource constraints in these countries are less severe, there is more scope for direct action to alleviate poverty by consumption-oriented measures such as a food subsidy program. In the low-income, late-developing countries, improving the well-being of the 60 to 80 percent of the population living in rural areas depends predominantly on the design and implementation of dispersal strategies. The first functional area relating to the distribution of land ownership or access to land is, therefore, fundamental.

Asset Distribution and Access

Land reform is an area in which the tendency to equate feasibility and desirability has been especially conspicuous. Despite an enormous and eloquent literature and much political discussion about the virtues of land reform, few countries have effectively implemented redistributive land reform. Postwar land reform programs in Japan, Taiwan, and South Korea are outstanding examples of effective redistributive programs not depending on a revolution. And, the success of the land reform programs in those three countries depended on special circumstances.

Partial success has been achieved in other countries. The postindependence land reforms in India drastically reduced the control over land by zamindars, the upper tier of landlords. But, ownership of land is still highly skewed toward the "big" and "large" farms. There was a fair amount of land redistribution in the early years of the martial-law regime of President Marcos in the Philippines. There was a rather surprising degree of success in converting share-tenancy arrangements into fixed rent leaseholds. Important categories of agricultural land were, however, excluded from the provisions of the land reform decree; actual implementation has been slow.

Economic and equity arguments for land reform are especially cogent in Latin America. As a result of the huge land grants of the colonial period, later acts of confiscation, debt peonage, and other processes, a large percentage of the agricultural land came under the control of extremely large haciendas, seldom smaller than 1,000 acres. In 1910 on the eve of Mexico's revolution, about 90 percent of the rural families in the country owned no land (Hansen, 1971, p. 147). Furthermore, the hacendados (owners of these enormous haciendas) frequently refused to rent any land to peasants. This is not surprising because these large haciendas combined a considerable monopoly of farmland with a monopsonistic position as employers of rural labor over an extended area. It was in the interest of hacendados to maintain their privileged position and to bind "their" peasants to the hacienda where they constituted a cheap and docile labor force.

Agricultural land in most countries of Latin America is extremely concentrated among few owners, accounting for so much attention to land reform. The political prospects for land reform are relatively good because the landlord class is so small that its political power must be considerably less than its economic power. According to Adams, (1973, p. 136) the prospects for redistributive land reform are probably very good in northeast Brazil because "120 families own most of the high-quality land."

We do not know why that expectation has not been more fully realized. However, three factors may have been important. First, because of the substantial industrialization and urbanization in countries such as Brazil, Colombia, and Mexico, their economies have become heavily dependent on the marketable surpluses produced on large, commercially oriented farms. Therefore, the industrial and rural elites oppose land redistribution for fear that production would drop and food for urban dwellers would become less available. Second, there has been little agricultural research and program experience in dispersal strategies under the rainfed conditions of Latin American agriculture. Third, a number of the regimes carrying out land reform have promoted collective farms rather than small but efficient family farms.

Schuh (1978) is among the few agricultural economists to challenge the general consensus succinctly stated by Adams (1973, p. 134): "Almost never does land reform decrease production, occasionally it has a neutral effect, most often it has a positive impact." Schuh's negative assessment of land reform is supported by the argument that Chilean land reform under Allende adversely affected agricultural production. Valdes (1974, pp. 410-13) argues

persuasively that the land reform hurts production
because collectivization as part of the reform package
weakens farmer incentives.

USF patterns of agricultural development in Japan,
Taiwan, and Korea were established long before the land
reform programs that were carried out following World
War II. In Taiwan in 1920, for example, 43 percent of
the farms registered owned less than 6 percent of the
land, and 11.5 percent of the owners had title to 62
percent of the land (Thorbecke, 1979, p. 137). The
size distribution of ownership units remained highly
skewed in all three countries until the postwar land
reforms. It was nevertheless possible to pursue
effective USF strategies for agricultural development
because operational units were uniformly small. Almost
without exception, large landowners found it profitable
to rent out their land to tenants who cultivated small
units very intensively, employing labor-using, capital-
saving technologies. It is easy to criticize this
situation from an equity point of view because the
tenant farmers were obliged to pay such "exorbitant"
rents that their net income was extremely meager.
However, the fact that tenants were prepared to enter
into such arrangements is clear evidence that expected
tenant income compared favorably with the prospective
returns from working for wages.

Economic advantages of cultivation by tenants in a
late-developing country are essentially the same as for
owner cultivators: they achieve fuller and more
efficient use of the relatively abundant supply of
rural labor because they rely primarily on family labor
in contrast with large farms which rely on hired
labor. Because large farms operate on commercial
principles and aim at profit-maximization, they do not
employ labor beyond the point where the marginal
product of labor is equal to the prevailing wage rate.

There are a number of theoretical considerations
which may account for "labor-market dualism" in
developing countries and the common tendency for the
effective (imputed) price of the nonwage labor on
family farms to be below the prevailing wage rate. 1/
Because of income-sharing among family members, the
supply price of labor is likely to be closer to the
average product of labor on a small farm than to its
marginal product. Moreover, because of uncertainty
about the availability of wage employment, small farm
units are likely to prefer to allocate their labor to
the intensive cultivation of their own rented land
because it provides a relatively assured annual income
even though the imputed price of their labor is below
the prevailing wage rates. To the extent that the
available supply of family labor is viewed as a fixed
cost, there is an incentive to adopt a cropping pattern

and labor-intensive techniques of cultivation and ancillary activities such as poultry rearing, milk production, or sericulture that expand the income-earning opportunities for the family work force. Family farms also provide enough flexibility for women to combine their child rearing and other household obligations with farming and auxiliary activities.

For these and other reasons, labor use on small farms will be more intensive than on large farms. More important than these theoretical arguments, however, is that small family farms invariably use more labor per acre cultivated than large farms. The USF patterns of agricultural development in Japan and Taiwan illustrate this tendency for small operational units to emphasize labor-using and capital- and land-saving innovations which are complementary to the relatively abundant labor resource. Survey data from contemporary developing countries also demonstrate more intensive use of labor on small farms. Equally important is the fact that yield per acre and total factor productivity tend to decline as farm size increases. Furthermore, a large body of empirical evidence indicates that these efficiency advantages of small family farms apply to share-tenancy as well as to owner-cultivators (Berry and Cline, 1979, p. 127).

Tenancy is often the most realistic option for realizing a USF pattern given the strong political opposition to redistributive land reform in many developing countries. It is only a "second-best" solution as compared to a redistributive land reform. Tenant farmers obviously do not benefit as much as owner-cultivators from increases in farm productivity and output because of the heavy rental payments that they are obliged to make to landowners in countries where there is a large farm population competing for a limited quantity of arable land.

The fact that landlords are able to demand such high rental payments has generated widespread support for "tenancy reform" (that is, the fixing of legal ceilings on land rental payments) as a substitute for a redistributive land reform. The political resistance to legislated ceilings on rental rates is less determined than the opposition to redistributive land policies. And, rental ceilings can be easily evaded. When a country's farm population is continuing to increase and alternative employment opportunities are severely limited, tenants and would-be tenants have a clear interest in colluding with landlords in ignoring the rental ceilings. Moreover, the enforcement of legal ceilings on rental payments requires continuing local surveillance where the political power of landlords is especially strong.

Pursuit of a desirable but infeasible goal adversely affects intended beneficiaries. If landlords simply ignore the legal ceilings on rent, the main effect of tenancy reform is to reduce the already limited bargaining power of tenants. Tenants relying on illegal, verbal agreements find it even more difficult than usual to obtain credit. A more common outcome of attempts at tenancy reform is for landlords to evict their tenants and convert large ownership units into large operational units (Lipton, 1978, p. 330). Landowners are then able to retain all the economic rent accruing to their land without having to be concerned about the legal restrictions on the rent collected from tenants. In addition, landlords may fear that tenants will establish a claim on their land which they may be able to make effective if the government begins a redistributive land reform.

Adverse effects on employment opportunities of setting up large operational units are related to another disadvantage of large farms, compared with small family farms. Creation of large operational units encourages adoption of inappropriately capital-intensive, labor-displacing technologies. The biological nature of agricultural production means that farming operations are spread out in time and in space. It is therefore much more difficult to supervise a large labor force than in a factory setting. Serious problems of shirking and poor work performance on the part of the hired workers arise. Furthermore, because of the high degree of variability that characterizes farming activities, there are numerous on-the-spot supervisory decisions needed from an individual performing routine tasks (Brewster, 1950). These characteristics of agricultural production fit decentralized decisionmaking. Small family farms, whether cultivated by tenants or owner-cultivators, have an important efficiency advantage: family members have an incentive to work hard and exercise judgment and initiative because they have a direct interest in maximizing production. Renting land to many small tenants is the most efficient way in which a large landowner can, in effect, "hire" labor to cultivate land. Such an arrangement uses the labor-intensive technologies appropriate when labor is relatively abundant and capital is scarce and expensive.

Large farm operators will minimize costs of supervision and problems of poor work performance by adopting mechanized techniques. Even though these relatively capital-intensive technologies are socially unprofitable because of the lack of alternative employment opportunities for the displaced labor, they are often privately profitable to individual farm

operators. The prevalence of low interest rate
policies and the underpricing of tractors noted in
chapter II obviously exaggerate the private
profitability of these inappropriate capital-intensive
technologies.

Two lines of thought have blocked USF patterns of
development. One line of thought has advocated large
operational units because of the alleged importance of
economies of scale. The other school of thought has
promoted policies and a climate of opinion hostile to
tenancy.

The first viewpoint is well illustrated by a statement by a former vice chancellor of India's Punjab
Agricultural University. "An efficient farm must have
a tractor and a tubewell," he claims, or "multiple
cropping and timely sowing is not a practical possibility." Therefore, he continues, "he must have a
minimum economic holding of 20-25 acres of irrigated
land." In his view, this "explodes the myth" that the
new technology of the seed-fertilizer revolution is
neutral to scale, a myth which, he asserts, "was
invented by those who wanted to promote a low ceiling
for landholdings, ignoring its evil effects on
production" (M. S. Randhawa as quoted in Franda, 1979,
p. 19).

The emphasis on "a minimum economic holding of
20-25 acres" ignores the structural-demographic
conditions which had reduced India's average farm size
to less than 4 acres by the early 1970s. Also perti-
nent is the technical and economic efficiency and the
record level of multiple cropping achieved in Taiwan
while farmers still relied on animal-draft power.
Under conditions prevailing in northern India and
Pakistan, a tubewell is often an important technical
complement to high-yielding varieties and increased
fertilizer use. But, tubewell water sales are fairly
common so that irrigation water is a divisible input,
although to a lesser extent than seeds and fertilizer.
Tractor mechanization, however, has very little positive effect on crop yields. Only atypically large
farms need tractors to facilitate multiple cropping.
2/ Moreover, economies of scale do become important
when tractor-powered technologies are adopted. In
addition, once the initial investment has been made in
a tractor there is a built-in incentive to adopt
increasingly labor-displacing technology as farm
operators acquire additional tractor-drawn implements
and the knowledge and skill to further mechanize. But
economies of scale are of very limited importance when
small-scale owner-cultivators or tenants use the
divisible, labor-intensive technologies appropriate to
a late-developing country.

The other line of thought has made it more difficult to achieve a USF pattern because it has created a policy environment hostile to tenancy, especially the crop-sharing tenancy so common in less developed countries. Share tenancy is regarded as "evil" because, it is claimed, it is an inefficient as well as an inequitable institution. It has only been during the past decade that the conventional view about the inefficiency of share tenancy has been challenged seriously. 3/ The earlier orthodox view was based on an incomplete analysis and applied a misleading analogy between a share tenancy contract and an ad valorem excise tax. The controversy continues. But a consensus has emerged that share tenancy does not often cause significant inefficiency in resource allocation. Most empirical studies of efficiency under owner cultivation and share cropping fail to reveal any significant differences in allocative efficiency. Some studies reveal an appreciable difference in the rate at which share croppers increase their use of fertilizer and other purchased inputs. But, that difference seems attributable to policies and institutional arrangements that put tenants at a disadvantage in terms of access to credit and other inputs. The experience of Japan and Taiwan demonstrates that this institutional bias can be avoided.

It has become common for landlords and tenants in India to share the cost of purchased inputs in areas where the high-yielding, fertilizer-responsive varieties have spread (Bardhan and Rudra, 1980). A USF pattern based on owner-cultivation is the preferred solution; but, even equitable distribution of land ownership will not significantly improve the income and well-being of farm people unless effective measures are taken to promote technological progress and accelerate expansion of employment opportunities within and outside agriculture. Success in increasing the opportunity cost of labor by expanding employment opportunities rapidly enough to tighten the labor supply/demand situation is also the most reliable way to improve the bargaining power of tenants and to increase their income and effective demand. Appropriate measures within the other six functional areas can all make significant contributions to that process. We will emphasize, for example, that land-saving innovations such as high-yield varieties and increased use of fertilizers tend to offset the adverse consequences of population pressure and land scarcity.

Forestry is a rural asset, access to which can augment the income of rural poor in many countries. This has indeed been achieved in several countries with FAO assistance. In Korea, for example, more than 1 million hectares of trees were planted through village

forestry cooperatives. These trees provided members
with fuelwood, timber, mushrooms, and oak leaves for
use and for sale. An FAO report (The State of Food and
Agriculture, 1981, p. 99) states that these forestry
projects effected a significant shift in resources from
the richer to the poorer members within project
villages through a profit-sharing scheme with the large
landowners. FAO supported this scheme by helping
develop the technical packages for the program and
training to government forestry staff.

FAO also reported that, in Mexico, Guetamala,
India, and the Philippines, forestry and other forest-
based activities have helped augment incomes of the
rural poor. As users of fuelwood for cooking and often
the main gatherers and sellers of forest products
(other than timber), women are affected by forestry
projects (FAO, ibid.).

Planning and Policy Analysis

The quality of the planning and policy analysis
that guides the design, redesign, and implementation of
agricultural development policies has a major influence
on the rate and pattern of agricultural development. A
basic requirement for achieving broad-based,
employment-oriented agricultural development is an
appropriate balance in the allocation of scarce
resources among the seven functional areas. But comprehensive economic planning as a major instrument for
achieving more rapid development has not been hugely
successful. The shortcomings of comprehensive planning
have been the result of the inevitable limitations of
intellectual cogitation or a "thinking through"
approach for resolving the ill-structured problems of
rural development. The cogitation approach to comprehensive economic planning seeks to anticipate all
possible outcomes, think through all possible contin-
gencies, and identify optimal plans for implementation. The overwhelming emphasis is on determining what
to do, to fix goals for growth of GNP, production
targets for individual crops, etc. The questions of
how to realize those goals tend to be regarded as mere
details. However, a "shift in focus to technological
and institutional details is long overdue. The most
serious problems lie, not in the grand design, but in
what has the superficial appearance of 'details'. 4/

Limitations of a "thinking through" approach to
planning and policy analysis are inevitable because of
the limitations of human cognitive capacities and the
scarcity of human attention. Nobel laureate Herbert
Simon has emphasized that "The dream of thinking every-
thing out before we act, of making certain we have all
the facts and know all the consequences, is a sick

Hamlet's dream" (Simon, 1971, p. 47). The intellectual cogitation approach to planning is also infeasible because it presumes the existence of agreed upon criteria of goodness of value by which alternative solutions can be judged (Lindblom, 1977, p. 322). W. Granger Morgan, one of the ablest practitioners of the art and craft of policy analysis, has defined "good policy analysis" as a systematic effort "to evaluate, or to order, and structure incomplete knowledge so as to allow decisions to be made with as complete an understanding as possible of the current state of knowledge, its limitations, and its implications" (Morgan, 1978, p. 971). In considering a problem as complex and ill-structured as rural development, one must consider a host of variables and changing interrelationships among those variables. This means that complete knowledge and understanding are impossible. In addition, decisions by policymakers will inevitably be influenced by subjective judgments, preferences, values, and vested interests. Morgan therefore emphasizes that good policy analysis should avoid drawing "hard conclusions unless they are warranted by unambiguous data or well-founded theoretical insight" (Morgan, 1978, p. 971).

In designing strategies for agricultural development, we usually do not have the "unambiguous data or well-founded theoretical insight" required for hard conclusions. Instead, we must contend not only with numerous interacting variables but also with the behavior of people and their interactions with each other and their environment. Because these interactions are mediated through a variety of organizations and institutions, it is essential that these entities be studied for their important role in the development process. Organizations and institutions rely on a variety of social techniques of "calculation and control" (Dahl and Lindblom, 1953). In this view, "organization" provides the framework for calculation and control through which collections of individuals attempt to determine what each should do and to assure that each does it. The techniques of calculation and control used include the techniques of exchange as in markets, the voting and bargaining of politics, and the hierarchical techniques of bureaucracies and large private organizations.

Because of the inevitable limitations of comprehensive economic planning, markets and prices represent a particularly important mechanism of social interaction which provides an essential complement to intellectual cogitation. A major feature of the process of economic development is the expansion and evolution of a variety of markets. In addition to the markets for food and other consumer goods and services, these include markets for an increasing variety of intermediate products such as steel and industrial chemicals and markets for labor, for land, for capital, and for financial services. This growth of markets is an essential feature of the division of labor and specialization among producers, the differentiation of function in specialized institutions such as universities and research institutes, and the growth of interindustry specialization as between agriculture and manufacturing. Those processes of specialization and the increased exchange and interdependence made possible by the growth of markets are in turn critical to the increases in productivity upon which economic progress depends (Johnston and Kilby, 1975, chapter 3; Kuznets, 1971).

Markets and prices play an especially critical role in facilitating agricultural development. We have stressed the importance of decentralized decisionmaking in agriculture. Market mechanisms and market-determined prices are a remarkably economical mechanism of calculation and control for guiding and harmonizing the decisions and actions of millions of producers and consumers. This is in part because a price system is such an efficient mechanism for transmitting information. The range of information that could be handled by alternative means of communication that are available or that could be created without excessive cost is very limited (Arrow, 1974, chapter 4).

Private firms in a reasonably competitive market can perform many economic functions satisfactorily with little or no government intervention. Requirements for calculation and control are, of course, most easily met with the simplest forms of organization. The great practical advantage of the family farm is that individual farm operators are able to make decentralized production decisions based on intimate knowledge of their resources and in response to price signals determined by market forces. Problems of incentive and work performance are minimal because the working family members are directly interested in the outcome of the enterprise. Even in large private firms, calculation and control are accomplished largely by responses to price signals that achieve profit maximization. Further, coordination problems are simplified because internal control and management in large private firms are usually hierarchical and simplified by the criterion of cost minimization. Private profit-maximizing firms are also relatively flexible and capable of making quick decisions in response to new opportunities or changing conditions.

A particularly important proposition, stressed by Galbraith, is that market mechanisms "economize on scarce and honest administrative talent." In contrast, reliance on government agencies or public sector firms means that "the greatest possible claim is placed on the scarcest possible resource. That is administrative talent, with its complementary requirements in expert knowledge, experience, and discipline" (Galbraith, 1979, p. 111).

This favorable view of the role of prices and markets is rejected in many developing countries, for many reasons. Sometimes this is a consequence of equating a market system with a laissez-faire economy, although few economists would argue that "getting prices right" is a sufficient basis for economic policy. Probably more important is the tendency to equate reliance on markets with "capitalism" as experienced when the economic policies of colonial powers often stifled industrial growth in their colonies.

Questions of equity, however, can form an important criticism of reliance on market-determined prices. This is especially likely to be the case in developing countries where the existing distribution of wealth, including human capital, is so unequal. The effective demand which interacts with supply to determine relative prices, allocation of resources, and distribution of income is a function of the existing income distribution. As a result, the "voting" by effective demand is highly unequal. Using the growth of aggregate national income as a measure of development "implies giving 10 to 20 times as much weight to a 1 percent increase in the incomes of the rich as to a 1 percent increase in those of the poor" (Chenery, 1980, p. 27; Ahluwalia and Chenery, 1974, pp. 39-42).

It is often assumed that relying on a system of government-administered prices rather than those determined by the market will enable a society to achieve a more equitable outcome. The problem is not that simple. Development experience of the past 25 years demonstrates that failure to use market-determined prices which accurately reflect the social opportunity cost of resources in some cases can have adverse effects on efficiency and on economic growth. Moreover, the price distortions introduced frequently exacerbate rural poverty even though the policies are ostensibly adopted for the benefit of poor farmers. We emphasized in chapter II that this is particularly likely to be true of low-interest rate policies and programs for subsidized distribution of fertilizer and other farm inputs. Such policies invariably create an excess demand situation which necessitates some form of administrative rationing. And, the relatively wealthy and politically powerful farmers are generally the principal beneficiaries of such policies.

Are we arguing in support of the Jeffersonian dictum that "that government is best which governs least"? Such is not the case. A program of redistributive land reform is likely to be highly desirable for economic as well as social reasons, subject to the important proviso that it is politically feasible to implement the program effectively. On the other hand, we urge great caution in advocating "tenancy reform" and also point to the danger of promoting the rhetoric of redistributive land reform without its implementation.

We do advocate that government policies and intervention programs should concentrate on activities important to the development process "which are not likely to be undertaken without planned public intervention" (Lele, 1975, p. 191; emphasis added). It is not only because administrative talent is scarce in late-developing countries that the opportunity cost of burdening the central administration with managing a system of government-administered prices is high. There is in fact a wide range of activities in which direct government action is indispensable because markets would perform poorly if at all without it. This applies particularly to the "public goods" which we emphasized in the introduction to this chapter. The activities that we will stress in analyzing the remaining functional areas are all of great importance because they enhance the production possibilities available to farmers. They foster increases in agricultural productivity and output and widespread increases in the incomes and effective demand of farm households.

Some of the most serious deficiencies of development planning stem from the fact that planners, because of their backgrounds and interests, spend most of their time in the capital city. This "capital fever" contributes to the urban bias which so often adversely affects expansion of agricultural production and the well-being of the farm population (Lipton, 1977). Moreover, programs for subsidizing agricultural credit and inputs frequently adopted to counter adverse effects of turning the terms of trade against agriculture usually benefit large farmers, thereby reinforcing a DSS pattern.

The concentration of decisionmaking and of the scarce resource represented by the analytical skills of planners and economists in the capital city also makes it difficult to design and redesign programs on the basis of an adequate understanding of realities at the village level. There is a need for greater decentralization of analysis and decisionmaking, including systematic learning from the actual implementation of agricultural development programs. We emphasize in the final section of this chapter that planners also need to give much more attention to the design and performance of the research, extension, and other organizations that are required in order to provide the "public goods" that enable farmers to increase their productivity and output. Those requirements are especially critical to the success of a broad-based, USF pattern of agricultural development.

Development and Diffusion of New Technologies

Agricultural research to generate a sequence of promising innovations is absolutely essential to USF patterns of development. In many instances, however, agricultural research and extension programs have failed to support widespread increases in farm productivity and output. We attempt to identify reasons for these shortcomings and the changes likely to increase the impact of agricultural research.

In Japan, Taiwan, and Korea, the development and diffusion of high-yielding, fertilizer-responsive varieties together with expanded use of organic and chemical fertilizers and the strengthening of irrigation and drainage systems constituted the all-important dispersal strategy. The overwhelming importance of rice in all three countries also facilitated the strategies. It took approximately a decade for agricultural scientists working in Taiwan to select adapted japonica rice varieties and to develop agronomic practices making it possible to grow them successfully under the low latitude, tropical conditions of Taiwan. The ponlai varieties were only being grown on an experimental basis in 1922. But, within a decade, nearly one-third of the total rice area was planted to those improved varieties. Much of Taiwan's rice area continued to be planted to the local indica varieties, and it was not until 1957 that a major breakthrough was made in developing semi-dwarf, fertilizer-responsive indica varieties with a yield potential similar to the japonica-type ponlai varieties.

The speed with which progress was made in developing and diffusing the high-yielding wheat and rice varieties was facilitated greatly by cumulative progress in agricultural science and by the important institutional innovation of a network of international agricultural research centers. An important example of the former was the identification of the genes that determine photoperiodism in wheat and rice which made it possible to develop varieties that can perform well over a wide range of latitude because they are insensitive to changes in daylight.

The spectacular success of the International Rice Research Institute in the Philippines in developing semi-dwarf varieties was made possible in part by the earlier breakthrough in Taiwan in developing high-yielding, fertilizer-responsive <u>indica</u> varieties. Similarly, the enormous success achieved in developing semi-dwarf, high-yield varieties of wheat in Mexico by the cooperative program of the Rockefeller Foundation and the Mexican Government and later by CIMMYT owed much to earlier research in Japan and in the U.S. Pacific Northwest.

There are four problem areas of crucial importance in assessing programs for the development and diffusion of new technology. First are the problems related to the "yield gap": the large difference between the yield levels obtained by the great majority of farmers as compared with the potential of the new varieties. The second concerns the need to further strengthen national research and extension programs, a problem which includes but is not limited to the need to increase budget support for these activities. A third problem relates to the special difficulties that need to be confronted in developing technological innovations that will be feasible and profitable for small farmers under rainfed conditions. The final problem relates to the disappointing impact of high-yielding varieties on employment opportunities in agriculture and on income distribution.

<u>High-Yielding Varieties and the "Yield Gap" Problem.</u> By 1976-77, 29 million hectares were being planted to high-yielding varieties (HYVs) of wheat in developing countries, accounting for approximately 44 percent of their total wheat area. In Asia alone, nearly 20 million hectares (72 percent of the wheat area) were planted to these varieties. High-yielding rice in the same year covered 25 million hectares. Virtually all of this area was in Asia, where approximately 30 percent of the total rice area was in high-yielding varieties (Pray, 1981, p. 71). The economic significance of the high-yielding varieties of wheat and rice derives from the fact that farmers obtain a larger increase in yield per kilogram of fertilizer. Even more important, these varieties can maintain favorable grain/fertilizer ratios with heavy fertilization. This latter characteristic is in notable contrast to traditional varieties which lodge (topple over) with high levels of fertilizer. The fact that improved seed-fertilizer combinations are a divisible as well as a highly profitable innovation has contributed to their rapid spread among small and large farmers.

Although the high-yielding varieties spread even more rapidly than had been anticipated, the increase in

farm-level yields has not reached expectations. 5/
Research workers at IRRI and other international
agricultural centers have been giving a great deal of
attention to this "yield gap" problem. Of crucial
importance are deficiencies in the availability and
management of irrigation water. One of the major
requirements for investment in rural infrastructure
that is considered in the next section relates to the
need to expand and improve facilities for irrigation
and drainage. An important part of the water supply
problem, however, relates to deficiencies in the
institutional arrangements and also mechanical devices
for insuring timely and equitable distribution of
water. The Technical Advisory Committee of the
Consultative Group on International Agricultural
Research accords a high priority to water management
problems (CGIAR, 1981, p. 55). The existence of the
yield gap has also helped focus attention on a more
general problem: research-extension systems in
developing countries often fail to obtain an adequate
understanding of the major constraints faced by
farmers. Therefore, many technological innovations
have relatively little positive impact on small farmers.

    National Research and Extension Programs. The
principal recent additions to agricultural research in
developing countries largely resulted from the dramatic
expansion of the so-called "CGIAR system." The consultative group now coordinates funding and operation
of nine international centers and four other CGIAR
institutions, including IFPRI (International Food
Policy Research Institute) and IBPGR (International
Board for Plant and Genetic Resources). 6/ In 1966,
the core operating expenditures and capital outlays for
the three centers then in existence amounted to less
than $2 million. In 1972, the system included six
international centers and the core expenditure
increased to $19.5 million plus $2.2 million for
special projects. By 1980, the total core expenditures
had increased to almost $124 million and an additional
$18 million was spent on specific projects. Part of
the increase was the result of price inflation, but the
average growth in core expenditure was close to 15
percent in real terms. The original funding of the
international centers was provided by the Ford and
Rockefeller Foundations, but many countries now
contribute to their support. In 1980, the United
States contributed $29 million, 25 percent of the
total, and the World Bank provided $12 million.

    The initial centers, IRRI and CIMMYT, have had the
greatest impact on agricultural production. This is
partly because of the enormous importance of rice and
wheat in the food economies of developing countries.
But, their success is also related to the fact that

those institutions drew upon a substantial amount of prior research. One estimate placed the internal rate of return of the research carried out at IRRI at an astounding 84 percent (Evenson, 1978, p. 238).

The international centers have enhanced productivity of national research programs and generated increased political and financial suppport for agricultural research. Both results are most evident in Asia. Evenson estimated that the internal rate of return of investments in agricultural research programs in Asia increased from 30 percent to 74 percent between 1965 and 1975 (Evenson, 1978, p. 238). Expenditures for agricultural research in Asia increased from $356 million in 1965 to $646 million in 1974 (in 1971 constant U.S. dollars). Over this same period, expenditures for agricultural research in Africa on a comparable basis increased only from $113.5 million to $141 million. Between 1965 and 1974, expenditures on agricultural research in Latin America increased by 133 percent from $73 to $170 million, an even greater increase than the 88-percent rise in the level of expenditure on research in Asia. However, the expenditure on agricultural research in Latin America still accounted for only 1.2 percent of the value of agricultural production compared with 1.85 percent in Asia (excluding China), 2.2 percent in Western Europe, and 2.7 percent in North America and Oceania (Boyce and Evenson, 1975, pp. 3, 8). The increase in agricultural research expenditures in Africa between 1965 and 1974 was a mere 0.24 percent, an outcome unquestionably related to the limited results achieved by agricultural research programs in that region.

National research programs in developing countries have also benefited greatly from training programs carried out by the international centers. The recent report of the CGIAR Review Committee has recommended that training priority should be given to the training of research leaders and managers and to maximizing the "multiplier effect" of their training activities. The scientists and administrators involved in the CGIAR system are keenly aware of the importance of empha- sizing the complementarities between the international centers and national research programs. They recog- nize, however, that there is no substitute for strong national agricultural research programs. Only strong national programs can undertake the adaptive research, as well as applied, strategic, and basic research, necessary to improve local productivity. Limited development of national capabilities for agricultural research will limit the ability of the developing countries to benefit as much as they might from the work of the international centers.

Tropical Africa is a special case with several reasons for the continuing weakness of national agricultural research programs. Political and budget support for research has not been reinforced by many success stories. In addition, political leaders and policymakers within the newly independent countries in Sub-Saharan Africa tend to be more preoccupied with political and distribution problems than with factors affecting productivity. Leonard has pointed out, for example, that during the colonial period in Kenya the European farmers exerted strong and effective pressures on the government to support a substantial agricultural research program. In 1963-64, just prior to independence, 17 percent of the development budget was devoted to research; in 1970-71, only 6 percent of the development budget was allocated to research (Leonard, 1977, p. 247). He attributes this to the tendency for African leaders to view development as primarily a question of access to resources and as a political problem. Consequently, issues such as support for agricultural research and the efficient use of resources have often received relatively limited attention. "The short time perspective" of policymakers in Africa has had similar consequences (Lele, 1981, p. 548). Countries of Sub-Saharan Africa confront some particularly difficult problems in creating agricultural research programs. Some of these difficulties stem from the fact that most are small, especially in terms of GNP and population, limiting the resources available for agricultural research. On the other hand, the great diversity in agroclimatic conditions in many of these countries means that agricultural research must deal with a large number of crops and research findings must be adapted to a great variety of environmental situations.

University-level education in agricultural sciences and related fields is limited and has constrained research progress in Sub-Saharan Africa. This is in marked contrast with India where a network of agricultural universities was established in the 1950s and 1960s with substantial economic and technical assistance from the United States. Progress made in strengthening those institutions and enlarging the availability of well-trained agricultural and social scientists has made a notable contribution to agricultural research and also to the capacity to manage agricultural development programs. Chapter III noted the strengthening of agricultural research in Malaysia. Considerable progress has also been made in the Philippines and a number of other Asian countries. However, the situation in some Asian countries, most notably in Nepal, is more similar to the situation in tropical Africa. Economic and technical assistance for institution-building programs in Africa began much later and has been much more limited than in most Asian countries. This has made it necessary to continue to rely to a substantial extent on expatriate agricultural scientists. The late start has also reduced the capacity for policy analysis and for the effective administration of agricultural development programs.

Special Problems of Rainfed Areas. The rapid spread of HYVs has been largely confined to areas with controlled irrigation. This is evident in the uneven spread of HYVs of rice. In Bangladesh, Burma, and Thailand, for example, the percentage of the rice area planted to HYVs was only 16, 5, and 6 percent, respectively, in 1973/74. This was due in part to the unavailability of improved varieties adapted to conditions of uncontrolled flooding and deep water that require the planting of "floating rice." Elsewhere, the HYVs are not suitable because of inadequate or unreliable rainfall. The spread of HYV wheat has also been concentrated primarily in irrigated regions where the risk of large variations in yields does not discourage farmers from investing in fertilizers and HYVs. In many countries, especially in Asia, there is still considerable scope for expanding the area served by irrigation and drainage facilities, promising increased area in high-yielding, fertilizer-responsive varieties.

In much of Africa, Asia and Latin America, however, the scope for expanding irrigated agriculture is much more limited because of technical and economic constraints. This is particularly true of most of Sub-Saharan Africa, where dryland rice accounts for 40 percent of total rice area (IRAT, 1984, p. 14). Only recently has there been a real recognition of the special problems of increasing productivity and output among small farmers in rainfed regions. There have been some notable successes such as the spread of hybrid maize in certain farming areas in Kenya as reported in chapter III. But those successes were in areas of high and reliable rainfall. The more prevalent situation is one in which the level of rainfall, its year-to-year variability, and its seasonal distribution limit significantly the increases that can be obtained simply by the introduction of improved seed-fertilizer combinations.

The challenge is greatest in the semiarid tropical regions where, as scientists at the International Crop Research Institute for the Semi-Arid Tropics (ICRISAT) have emphasized, "more than 500 million of the poorest people in the world eke out a livelihood from the meager resources of land and capital in an unfriendly climate" (Ghodake et al., 1978, p. 1). Achieving progress under rainfed conditions usually confronts the difficult problems of making concurrent progress in developing high-yield, fertilizer-reponsive varieties and developing and introducing tillage and equipment innovations to improve soil and water management and to increase the timeliness of planting and weeding.

Major progress has been made in some developed countries--notably Australia, the United States, and Israel--in evolving technological innovations rewarding even under semiarid conditions. These technologies are, however, geared to heavy and costly tractor-powered equipment not feasible for small and poor farmers. For low-income developing countries, the greatest promise lies in simple, inexpensive, but well-designed animal-powered implements.

Farming systems research at ICRISAT has concentrated on animal-powered equipment, based on the recognition that it will be many years before most farmers in the semiarid tropics can afford tractors and tractor-drawn equipment. 7/ In less than a decade, ICRISAT research has demonstrated that large increases in net returns are possible with a combination of biological-chemical and equipment-tillage innovations (Ryan et al., 1980). This experience also illustrates, however, another significant problem confronting research on rainfed farming: the great local variation in the moisture-retaining capacity of different soils as well as rainfall.

There is a particularly urgent need for tillage and equipment innovations adapted to local conditions in late-developing countries, especially those in tropical Africa. About 80 percent of cultivation in Africa still depends on human labor and the hoe; numerous efforts to make the transition directly from the hoe to tractors have failed to benefit more than a small percentage of African cultivators. Furthermore, the growth of demand for simple and inexpensive equipment offers one of the most promising ways to foster the growth of rural-based manufacturing firms in Africa. Because of the limited development of manufacturing in these countries, tractors and tractor-drawn implements must either be imported or "manufactured" in one or two urban-based factories which merely assemble imported components. The value added and employment generated by such factories is limited. In contrast, small- and medium-scale machine shops producing simple implements can employ technologies that make considerably greater use of labor and local materials and thereby minimize requirements for capital and foreign exchange. Moreover, the evolutionary growth of such firms provides a valuable training ground for developing the technical and managerial skills needed for widespread and healthy growth of a domestic manufacturing industry.

Progress in realizing that potential has been limited. One obvious problem has been the lack of financial and manpower resources for all types of agricultural research. Another factor has been the tendency to question the need for publicly supported R&D on equipment innovations because, it is claimed, "the development of new machinery has generally been met by manufacturers" (CGIAR, 1981, p. 56). Such development did occur in the United States, many other developed countries, and to a considerable extent in a country such as India where there has been considerable development of rural-based manufacturing firms. But, private enterprise development has not been responsive, especially in tropical Africa, because of the chicken and egg problem: the need to accelerate progress simultaneously in identifying and diffusing well-adapted equipment and in promoting growth of local manufacturers to respond to demand which cannot become effective until appropriate equipment and tillage innovations have been identified and introduced to farmers. Critics of strengthening R&D efforts also point to the limited success of past programs to develop and test equipment, for example the Tanzania Agricultural Machinery Testing Unit mentioned in chapter III.

The agricultural engineering division at IRRI had considerable success in fostering local manufacture and expanded use by farmers of items such as small threshers and a simplified version of power tillers adopted widely in Japan in the 1950s and in Taiwan some 10 to 15 years later. The IRRI experience demonstrates the need for a sustained effort and a systematic method for evaluating promising areas for product development An agricultural economist working with that program further stresses the importance of giving explicit attention to promoting backward linkages through local manufacturing (Duff, 1980). Progress in developing and diffusing equipment and tillage innovations in rainfed regions depends on a "systems approach" involving engineers, agronomists, and other scientists in order to examine interacting effects on crop yields of equipment design, tillage methods, soil character-istics, moisture conditions, varietal improvement, and fertilizer use (Johnston, 1981).

Difficulties encountered in promoting wider and more efficient use of improved farm equipment points to a problem applying to many aspects of the research-extension system. The conventional model of agricultural extension assumes that extension field staff will ensure two-way communication between research workers and farmers. This has not worked well in developing countries. Even the link between research workers and extension staff has often been weak. An interesting feature of the training and visit system of organizing agricultural extension pioneered by Daniel Benor is its emphasis on organizational structures and management procedures to strengthen the linkages between research and extension staff (Benor and Harrison, 1977).

A related requirement also addressed by the training and visit system is to insure that the innovations developed by research and promoted by extension are feasible and worthwhile at the farm level. For example, efforts by agricultural extension staff to promote wider use of improved equipment and a program to foster its local manufacture should be preceded by farmer acceptance trials to verify that farmers find the implements useful. A more general technique involves use of reconnaissance surveys of agrotechnical and socioeconomic conditions to identify problems and possibilities that characterize farming systems in different agricultural regions. Such surveys can assist in determining research objectives and priorities. Subsequent trials carried out by research staff on farmers' fields help verify suitability of promising innovations. Tests or acceptance trials by representative farmers provide additional verification. A farming systems approach to adaptive onfarm research is similar and also promising (see Hildebrand, 1976; CIMMYT, 1980).

Employment and Income Distribution Effects. The defining characteristic of dispersal strategies and of a USF pattern is that technological progress has a widespread impact in improving employment opportunities and income distribution. It has often been emphasized, however, that the Green Revolution in India and other Asian countries has been associated with a deterioration in the employment situation and increased inequality in the distribution of income.

Moreover, critics often claim that those effects are inevitable consequences of an agricultural strategy that promotes expanded use of HYVs and fertilizers. The Green Revolution has not fulfilled the earlier highly optimistic expectations. But, this outcome should not be attributed to the nature of the core technology of improved seed-fertilizer combinations and the related emphasis on expanding and improving irrigation. A major factor contributing to the failure to improve the employment situation and income distribution is the continuing and rapid growth of the farm population and labor force. In countries such as India, Bangladesh, and Pakistan, a very elastic supply of labor has been associated with a very inelastic supply of farmland. This has prevented any significant tightening of the labor supply/demand situation so that most of the gains of technological progress have accrued to landowners as increased economic rent.

The employment situation and income distribution have also been aggravated by premature tractor mechanization and by the displacement of tenants as large landowners have gone in for direct cultivation. Those tendencies have been encouraged by ill-advised economic policies, including low-interest-rate policies and trade policies that have artificially reduced the price of capital, of foreign exchange, and the import price of tractors and other capital-intensive inputs. Moreover, the considerable concentration of effective demand on purchases of labor-displacing inputs such as tractors has tended to preempt the growth of demand for the simple types of farm equipment which otherwise would have stimulated more rapid growth of output and especially of employment in rural manufacturing firms.

Analysis of these issues has been bedeviled by a polarization of the debate between those who emphasize the importance of accelerating the rate of technological change and others who tend to condemn any technological innovations that reduce the demand for labor. We stress the need to be concerned with both the rate and bias of technological change. Technical progress leading to increases in productivity are absolutely essential for achieving rapid increases in agricultural output where average incomes are extremely low and poverty is not a matter of isolated pockets but a pervasive problem affecting about 30 to 40 percent of the rural population.

If the basis of technical change is in a labor-saving, capital-using direction, the majority of a low-income country's farm population will not benefit from gains in productivity and income. Evidence from India demonstrates that the growth of employment opportunities was not rapid enough to absorb the annual additions to the work force (Government of India, 1978, Chapter 4; Krishna, 1972, 1973). This underscores the importance of economic policies and a USF pattern of agricultural development to ensure that the growth of demand for labor exceeds the rate of increase of the rural labor force. Capital-intensive technologies displace more labor than "appropriate" technologies. Such technologies waste capital in comparison with labor-using, capital-saving technologies that are economically more efficient when labor is abundant and capital is scarce. The social cost of this use of scarce capital is very high; it deprives the economy of alternative possibilities for creating new jobs and achieving more rapid as well as more widespread economic growth.

But equity concerns, based on a narrow, short-run view of the effects of technical change can be disastrous. In Java, for example, the farm population is so large relative to the supply of land and other factors of production that even the shift from harvesting rice by the traditional ani-ani knife to the sickle is sometimes condemned because it reduces the demand for labor. But a policy that would seek to preserve employment opportunities by preventing any technological progress perpetuates poverty. We have stressed that HYVs normally increase labor requirements per acre when their adoption is not accompanied by excessively labor-displacing mechanization. But even this labor-using, capital-saving innovation reduces labor requirements per kilogram of rice or wheat produced. Otherwise, it would not be so attractive to farmers or so socially profitable. Technical progress is critical for the elimination of poverty because appropriate innovations release resources for production of new goods and services as well as expanded output of existing products. There are, however, likely to be situations in which the economic adjustments that lead to the creation of new employment opportunities may operate too slowly to prevent the emergence of serious short-run problems. To respond to those problems by stifling technical progress is self-defeating for a low-income developing country. An alternative that does merit serious consideration-- rural works programs to compensate for job loss due to technical change--receives attention in the next section.

Investment in Rural Infrastructure

In several of our case studies, we noted the key role of improvements in the rural road network and other communication facilities in fostering widespread increases in farm output. Investments in rural electrification can also facilitate the expansion of rural employment opportunities, especially if the pattern of agricultural development favors rapid expansion of rural-based manufacturing firms utilizing labor-using, capital-saving technologies. Frequently, infrastructure investments in irrigation and drainage are critically important to enlarge opportunities for productive employment in agriculture as well as to increase crop yields and output.

The historical evidence of Japan, Taiwan, and Korea provides evidence on contributions of improved water control to yield-increasing innovations for individual crops and to facilitate multiple cropping. We mention- ed in chapter II that underemployment of the agricul- tural labor force in Taiwan was reduced substantially between 1911-15 and 1956-60 in spite of a 50 percent increase in the farm labor force and a considerable reduction in the amount of cultivated land available per farm worker. An increase in the index of cropping intensity from 116 to 180 over that 45-year period was undoubtedly the most important single factor accounting for the expansion of employment opportunities.

A recent analysis of the effects of expanded irrigation and increased multiple cropping in India demonstrates that since the 1950s expansion of irrigation and increased multiple cropping have made a notable contribution to the growth of agricultural output. This contribution has become increasingly important as the scope for expanding the cultivated area has declined. Between 1950/51 and 1960/61 the "net sown" area in India increased at an annual rate of 1.16 percent; between 1970/71 and 1975/76 the net sown area increased at a rate of only .12 percent whereas the "gross cropped area" increased at a .5-percent rate because of expanded multiple cropping. Fortunately, there is still considerable scope for expanding irrigation and multiple cropping in India: India's index of multiple cropping in 1975/76 was 120, far below the peak level of 120 reached in Taiwan in 1966 (Narain and Roy, 1980, pp. 9, 10; Republic of China, 1974, p. 48). Considerable opportunities for expansion of irrigation exist in a number of other Asian countries, although the cost per hectare of command area is expected to rise because most of the lower cost sites for irrigation schemes have already been developed. According to an IFPRI study of investment and input requirements for accelerating food production in low-income countries between 1975 and 1990, one-half of the projected increase will come from the expansion of the area under irrigation (Oram et al., 1979).

During the 1970s there was a tendency to curb development assistance for irrigation and other infrastructure projects on the grounds that such investments do not confer direct benefits on the rural poor. This is another example of how a narrow, short-run view of the problem of eliminating rural poverty can be harmful to the intended beneficiaries. The effects of investments in infrastructure will have limited impact on employment opportunities and on increasing the effective demand of the great majority of the rural population if those investments are concentrated in areas dominated by large-scale, capital-intensive farm units. The concentration of investments in irrigation in northern Mexico is an important case in point.

A number of irrigation projects in Sub-Saharan Africa have been very costly in relation to the number of farmers who have benefited from those investments. The relatively limited scope for economically and socially profitable investments in irrigation projects in Sub-Saharan Africa is in part a consequence of a

physical environment which offers relatively few
low-cost sites for irrigation projects. There are a
few examples of successful irrigation projects, most
notably the Gezira Project in the Sudan. That scheme
owes its success, however, to soil and other conditions
suitable to the production of long-staple cotton, a
high-value crop. The same is true of the Mwea
irrigation project for rice in Kenya. A number of
other projects, such as the Office du Niger in Mali,
have been costly failures. The opportunity cost of
ill-advised irrigation projects in tropical Africa is
high; they divert the scarce resources of capital,
trained manpower, and the attention of policy makers
away from more extensive and profitable opportunities
that exist for increasing productivity and output under
rainfed conditions. Attention needs to be given to
accumulating hydrological data and developing technical
expertise and experience in the design, construction,
and management of irrigation projects. In general,
however, irrigated agriculture is too expensive and too
technically demanding to merit a high priority in the
countries of tropical Africa because of the greater
benefits that can be derived from expanding cultivation
and raising yields of rainfed crops. Small-scale
irrigation projects to permit year-round cultivation of
vegetables and fruits are a significant exception. The
economic and nutritional value of these crops is high.
Moreover, it is often possible to design schemes for
supplying water to villages for household use,
justified for health reasons and for reducing the time
and drudgery of hauling water, in such a way that the
supply is sufficient for watering small vegetable plots
and fruit trees.

A more appropriate response to the problem of rural
poverty has been to give greater attention to
exploiting the employment-generating potential of
investments in rural infrastructure projects. This has
been especially true of projects for the construction
of rural roads using labor-intensive methods. A major
project for building rural access roads in Kenya is of
considerable interest because it "proved that labor-
intensive methods are economical and technically
viable" (de Veen, 1979, p. 1). That study and a paper
by Judith Tendler (1979a) emphasize, however, that
organizational and management techniques need to be
modified considerably for the use of labor-intensive
methods rather than the capital-intensive technologies
on road-building equipment.

Recipient governments and especially their highway
departments are often reluctant to adopt labor-
intensive techniques. In addition, AID and World Bank
procedures have tended to encourage equipment-based
techniques. They are less demanding of staff time for

project preparation, and it has usually been easier to finance the foreign exchange costs of road-building equipment than local wage costs. AID, the World Bank, and ILO have in recent years sponsored research and seminars that have emphasized that labor-intensive techniques have the social advantage of generating more employment and are also more efficient given the low wage levels that prevail in many developing countries. The need to modify organizational arrangements and management procedures in order to realize the potential for labor-intensive construction of rural roads and other types of infrastructure emphasizes the importance of giving explicit attention to the organizational issues which we examine in the final section of this chapter.

Rural electrification is another type of infrastructure investment with potential contribution to the expansion of employment opportunities. In this case the potential for creating direct employment opportunities through labor-intensive construction methods is limited. The indirect effects of rural electrification on the expansion of job opportunities may, however, be significant (Tendler, 1980b). Tubewells and low-lift pumps powered by electric motors often have advantages over diesel engines, and we have seen that improved water control often permits a considerable expansion of opportunities for productive employment in agriculture. Growth of job opportunities in rural-based light industries can also be fostered by the availability of electric power. Research in the Philippines, for example, indicates that under favorable circumstances the increase in nonfarm employment that is fostered by rural electrification can be substantial. Whether such possibilities will in fact be significant depends on the influence of macro-economic policies and on the rate of growth of effective demand in rural areas. We discuss those issues below.

Attention has been given to labor-intensive rural works programs undertaken for the twofold purpose of providing supplementary income-earning opportunities for the most disadvantaged rural families and for improving the rural infrastructure. Many of these projects have been "food for work" projects that have utilized grain made available under P.L. 480 or the World Food Program. The Maharashtra Employment Guarantee Scheme initiated in 1972 has attracted considerable attention as a well-conceived and reasonably well-managed rural works program. The scheme emphasized the construction of infrastructure which enlarges the productive capacity of the state's rural economy, and it has been backed up by well-prepared, detailed plans for labor-intensive

construction projects. The Maharashtra scheme made a
substantial contribution to reducing unemployment and
underemployment in the state, and in the last few years
similar schemes have been launched in other parts of
India utilizing the large stocks of grain accumulated
by the Government of India. Maharashtra is one of the
more developed and wealthier states in India, and in
other regions the design and implementation of rural
works programs is more difficult.

In general, governments in developing countries
tend to be reluctant to commit themselves to financing
rural works programs because of budget constraints.
Clearly, rural works programs are not a substitute for
policies and programs to foster a USF pattern of
agricultural development with its important direct and
indirect effects in expanding employment opportunities. Such schemes, however, can be a significant
supplement. They can play a significant role in times
of drought or other natural disasters that reduce farm
employment opportunities and incomes. Again evidence
from India is relevant. According to a study of the
various mechanisms available to rural households for
coping with the severe problems that rise at times of
drought, the supplementary income from rural works
projects was the only one without serious adverse
effects on the human and physical capital of poor rural
households (Jodha, 1978; Binswanger et al., 1980).

Marketing and Storage, Input Supply, and Credit

The institutions and facilities required for
marketing and storage of agricultural products, for the
supply of inputs, and for the provision of credit are
essential elements of the rural infrastructure
(Peterson, 1981; Wharton, 1967). For these activities,
however, the "institutional infrastructure" is more
important than the "physical infrastructure." The
direct employment generated in carrying out these
functions are of limited importance, although in a
number of developing countries employment in "commerce"
bulks large in nonfarm employment in rural areas (World
Bank, 1978, p. 24). However, the efficiency with which
farm products are marketed and stored, inputs are
distributed, and credit is supplied to farmers has a
decisive impact on the rate of growth of agricultural
production. Moreover, the nature of the policies and
organizations that determine how those functions are
performed are important in determining the prospects
for successful implementation of dispersal strategies
and the realization of a USF pattern of agricultural
development.

There is no need to elaborate on our earlier comments about the effects of low-interest-rate policies and of subsidized distribution of farm inputs in strengthening the forces that lead to a DSS pattern of agricultural development. There is a good deal to be said about the choice of organizations for performing those functions. During the past three decades cooperatives and parastatals (quasi-governmental organizations) have frequently been the preferred types of organizations for performing those functions. When those organizations are ineffective in reaching small farmers, governments are likely to favor large farms and sometimes state farms because it is easier for parastatal marketing organizations to purchase the marketable surplus from large farms. The last section of this chapter deals explicitly with questions of organizational design and performance.

Rural Industry and Ancillary Activities

In recent years increased attention has been given to rural nonfarm activities as a source of employment and income in developing countries. A 1975 book, Agricultural and Structural Transformation by Johnston and Kilby, emphasized the interrelationships between agricultural and industrial growth. The empirical evidence available at the time was limited. Consequently, their study focused mainly on evidence that the authors were able to assemble pertaining to the experience of Taiwan, India, and Pakistan, together with Kilby's previous studies of small-scale industry in tropical Africa and his research on entrepreneurship in developing countries (Kilby, 1969, 1971).

A subsequent World Bank publication, Rural Enterprise and Nonfarm Employment, and a "state of the art" paper by Chuta and Liedholm provide additional evidence concerning the extent and nature of rural nonfarm activities. According to data for 18 developing countries, some 20 to 25 percent of the rural labor force is employed primarily in rural nonfarm activities. The corresponding figure for Guatemala, however, was only 14 percent whereas in Taiwan 49 percent of the rural labor force was in the nonfarm sector. In addition, rural nonfarm activities are usually an important source of secondary employment, often on a part-time or seasonal basis. It is, therefore, not surprising to find that nonfarm income accounts for a significant share of the total income of rural households--between 20 and 30 percent for villages studied in Pakistan and Nigeria, 36 percent in Sierra Leone, and 43 percent of the considerably higher income of rural households in Taiwan. It is also noteworthy that the share of nonfarm income appears to be especially large for rural households with little or no farm land (Chuta and Liedholm, 1979, pp. 4-7, 15).

A detailed examination of ancillary activities carried on by farm households and of the characteristics of rural nonfarm activities and of the factors that influence their growth is outside the scope of this book. It is necessary, however, to note their importance and to emphasize the major influence of both the rate and pattern of agricultural development on the expansion of output and employment in the rural industries.

The rate of growth of output and especially of the monetary income of the agricultural sector are clearly major determinants of the growth of rural nonfarm activities. Agriculture affects the growth of the nonfarm sector through backward and forward production linkages and final demand linkages. Forward linkages such as those related to milling grain and processing other agricultural products are most directly induced by the growth of agricultural output. Backward linkages related to the demand for farm inputs are influenced strongly by the rate of growth of farm output and incomes, but the nature of those linkages is also very dependent on the pattern of agricultural development. The final demand linkages which influence the growth of rural demand for consumer goods and services will also be very different under a USF pattern of development as compared with a DSS pattern.

Expenditures for chemical fertilizers are generally the most important category of farm production expenditure, but the backward linkage effects associated with the growth of demand for farm equipment are much more significant. The level and type of demand for fertilizer are not very sensitive to the type of agricultural strategy pursued, and the continuous process industries used for their manufacture offer little scope for adjusting factor proportions. The principal policy issue is whether to import or to manufacture fertilizers domestically; and a USF pattern of agricultural development will be favored by the policy that makes fertilizers available to farmers at the lowest possible cost.

In contrast, the choice between a USF and DSS pattern of agricultural development has a powerful influence on the type of farm equipment produced, on the technologies used in their manufacture, and on the effects of those backward linkages on the growth of domestic manufacturing. These differential effects apply not only to the type of products that receive a direct stimulus from increased demand by farmers, but also to the nature of the technological spillovers that influence the development of technical and managerial skills and of manufacturing capabilities.

We stressed in an earlier section that the demand for animal-powered implements and other relatively simple and inexpensive implements that is fostered by dispersal strategies favors the growth of small- and medium-scale machine shops, many of which will be located in rural towns and small cities. This was contrasted with the effects of a focus strategy which leads to a growth of demand among a subsector of large farms for tractors and other sophisticated items that will either be imported or assembled in large factories in metropolitan areas. In addition to being capital- and import-intensive, the scale and complexity of the technologies in those urban-based factories is much greater than in existing artisanal and manufacturing activities; as a result, they remain isolated enclaves with little impact in spreading technical knowledge and raising productivity.

The pattern of demand for farm equipment associated with dispersal strategies, however, fosters the emergence and development of machine shops and light engineering firms that concentrate initially on very simple techniques and products. Because the technologies are not dramatically different from activities already mastered by local artisans and technicians, their diffusion is likely to be fairly rapid. At the same time, the more experienced and competent firms can be expected to upgrade their technologies and products. This is partly in response to demand from farmers for new and moderately sophisticated implements such as seed-fertilizer drills or low-lift pumps. The strengthening and diffusion of skills in metalworking that results from this process is significant because those skills are important to many types of manufacturing. This includes a growing capacity to manufacture capital goods, which capacity is crucial for adapting imported technologies to local conditions and making them widely available to producers throughout the economy.

Government-supported R&D activities can, as we suggested earlier, foster expanded use of farm equipment appropriate to the requirements of small farmers with limited cash income and also facilitate the growth of local manufacture of such equipment. The creation of industrial estates and various other techniques are available for promoting the establishment of small- and medium-scale enterprises and the progressive upgrading of their technical and managerial skills.

Government policies have often obstructed the growth of "informal sector" manufacturing firms. This has been in part a consequence of their preferential treatment of the larger firms in the "modern sector." This includes loans at artificially low interest rates, *farms can grow w/t acreage increase!!*

allocations of foreign exchange, import licenses, allocations of scarce materials at official prices, tax holidays, and other types of preferential treatment that are usually not available to the small- and medium-size firms. These policy issues are outside the scope of agricultural policy, but they are significant obstacles to the achievement of broad-based, employment-oriented agricultural development and rapid growth of nonfarm employment opportunities.

Ancillary activities such as poultry-rearing, keeping a cow or two for milk production, or seri-culture can provide significant opportunities for productive employment and additional income for small family farms. India's experience since the creation of the National Dairy Development Board (NDDB) in 1965 has demonstrated the importance of animal husbandry and milk production "as a major instrument of social change, for supplementing the income and providing a large scope for employment" to small and marginal farmers (Government of India, 1978, p. 144).

At the end of 1976 over 4,500 village cooperatives with a membership of some two million farmers had been organized, and the number of small farms participating in the NDDB program has continued to increase. Milk and milk products were estimated to reach 35.5 million tons in 1982-83 compared with a level of 23.2 million tons in 1973-74. The village cooperatives promoted by the NDDB are a major element in this expansion. It is reported "that the program operates with a high level of efficiency and lack of corruption, and provides major social and economic benefits to the poorest members of the member village while assuring urban consumers a regular supply of quality milk products at fair prices" (Korten, 1980, p. 485). The organization-al aspects of the program are of great interest and we examine them in some detail in the following section. Kenya is another example of a country where the expansion of milk production by small-scale farmers has made a significant contribution to the incomes of small farmers. As we noted in chapter III, this is based on the availability of high-yield milk cows and requires careful management. In both cases the successful production of milk requires support from a competent veterinary service for artificial insemination and animal health.

Institutional Development: Improving Organizational Structures and Managerial Procedures

Development economists and planners have given relatively little attention to the organizational aspects of development. It has become increasingly clear, however, that the wide gap between ambitious plans for agricultural development and the results achieved is in large measure a result of inappropriate and ineffectual organizations. These issues related to organizational design and performance are especially critical to the implementation of a USF model of agricultural development because of the need to reach and actively involve large numbers of small-scale farm units.

A 1966 monograph by S. C. Hsieh and T. H. Lee argues that "the main secret of Taiwan's development" was "her ability to meet the organizational requirements" for agricultural development (Hsieh and Lee, 1966, pp. 103, 106). We emphasized earlier that organizational issues are especially critical in providing the public goods which will not be made available at socially optimal levels without interventions by publicly-supported organizations: institutions for agricultural research, extension, and for the construction and management of irrigation facilities and other types of infrastructure. The East Asian countries of Japan, South Korea, and the People's Republic of China as well as Taiwan are notable in the attention that has been given to creating a network of organizations to provide those support services.

Organization and development--This neglect of problems of organization is no doubt related to the tendency of planners and economists to abstract from such "details." It also reflects a view that organizational choices are unimportant or that they are not a subject which governments can do something about. But attempts to change what things are done usually must be accompanied by appropriate changes in how things are done. The recent efforts to introduce labor-intensive techniques of road construction are, as we noted above, a good example.

Fortunately, there has been a considerable increase in research on issues of organization and management during the past decade; it is clear that these issues are subject to experimentation and that well-considered government action can make a difference. 8/ Efforts to influence the design and management of development organizations must be guided by an understanding of the constraints and opportunities associated with organizational choices.

Research and analysis of organizational problems can lead to useful generalizations, but it is not possible to do justice to those generalizations in a brief summary treatment. In any country there is a wide range and variety of existing organizational structures. The problems which arise in connection with small community organizations at the local level are different from those of "facilitator organizations," which link rural people and their local organizations with a regional or national government. The difficulty of summarizing these issues is also a consequence of the importance of the historical dimension of social organization: organizations are organic entities which develop over time. Strengthening organizational capacity involves visible changes in organizational structure and a growth in administrative skills. It also involves less evident changes in cultural values and expectations: an existing organization represents a complex, dynamic system of linkages which is only partly understood by participants and observers.

Redesigning and reorganizing existing patterns of linkages is a tricky business. Any notion that policymakers and analysts have a "clean slate" on which to write their organizational prescriptions is naive. Drastic reorganizations are likely to damage existing capacities to resolve social problems. What is needed in most cases is a sequence of steps to mitigate the more serious failings of existing organizations. The practice followed by some aid donors, especially the World Bank, of creating a special project management unit is no doubt convenient for the donor agency. But for the developing country, it exacerbates the problem of strengthening its capacity to administer development programs.

Organizing the facilitators--Facilitator organizations have certain common features because of their reliance on a hierarchy of full-time bureaucrats. It is useful to make a distinction between facilitator organizations concerned with "managing the predictable" and those organizations which need to be especially concerned with "managing a learning sequence." The highly structured approach to reorganizing agricultural extension programs according to the "Training and Visit" system that we discussed above, is a case of "managing the predictable" because of the extension experience that has been acquired in administering agricultural extension programs in developing as well as developed countries. However, evolving effective organizational arrangements and managerial procedures for identifying and diffusing farm equipment innovations and for promoting their local manufacture must cope with the task of managing a learning sequence because of the lack of relevant experience and our limited understanding of how to attain those objectives.

Participation and the organization of local groups. It has become fashionable in recent years to emphasize the importance of participation and the creation of local organizations at the village level. It is desirable to tap local knowledge and understanding and to actively involve rural people in the design and implementation of programs intended for their benefit.

There is also a need for vertical linkages which tie local people and their organizations into the larger administrative and economic structure.

Most of the discussion of participation, however, has not gone much beyond the level of rhetoric. There is a tendency to treat participation as a free good, desirable in unlimited quantities. This ignores the fact that participation in a local organization requires an investment of time and energy of the participants. This is an investment which cannot be commanded by policymakers or administrators. It can only be induced.

Clark has suggested that the ability to induce effective participation will depend primarily on three features of local organizations. The first feature relates to the attractiveness of the benefits that are expected to result from participation. Local people cannot be expected to invest their time and energy in a local organization unless it offers desired benefits which are not otherwise obtainable at similar cost. The second feature is a requirement for at least a minimum degree of harmony among participants in their views as to the objectives of the organization and the means of reaching those objectives. The third feature concerns the simplicity of the techniques of calculation and control that are utilized.

All three of those features are important in influencing the feasibility and effectiveness of efforts to create local organizations to help build local problem-solving capabilities and to promote agricultural development. The third feature relating to simplicity of technique is especially relevant to several important and controversial issues of agricultural policy. One of these, the nature and role of agricultural cooperatives, will be considered later. The other issue concerns group or collective farming.

Two characteristics of an organization will largely determine the simplicity of the techniques of calculation and control that can be used. The first is simply the size of the organization: the larger the organization the more complex and difficult is the task of determining individual responsibilities and the distribution of benefits. The second factor relates to what Clark refers to as "communality." Whenever participants contribute their labor, land, or other resources to a common productive activity, calculation and control are inevitably more complex and more difficult than when each individual's contribution is reflected directly in his reward. We emphasized earlier that the great practical advantage of the individual family farm is that the cultivator and members of his family have a clear-cut incentive to invest their labor, knowledge, initiative, and skill in the family enterprise because of their direct interest in the outcome of their efforts. Some governments may nevertheless prefer the collective approach to agricultural production. Such decision, however, should at least be made with an awareness of the special organizational and incentive problems this entails.

Moreover, the success or failure of such efforts will be determined much more by the attitudes and response of participants than by the preferences and predilections of policymakers. The attitudes and responses of participants will depend in large measure on their perception of the balance between costs and benefits of the organizational effort. For reasons that were examined earlier in the chapter, economies of scale are not of much importance in agriculture, especially when a country's resource endowment makes it appropriate to emphasize labor-using, capital-saving technologies. On the other hand, economies of scale are clearly of major importance in the design, construction, and management of an irrigation system. Consequently, farmers are much more likely to invest their participation in an irrigation association than in a collective established to carry out crop cultivation in common.

Recent experience in the Philippines is of interest with respect to participation by local organizations. B. Bagadion and F. Korten (1980) have provided an account of how the National Irrigation Administration has approached the task of <u>learning how</u> to organize viable local organizations for managing small-scale irrigation systems. These so-called "communal" systems, in contrast to "national" systems, serve approximately half the irrigated land in the Philippines. The way in which pilot projects and a "learning laboratory" approach were used to involve farmers into the project planning process for constructing or improving and managing these small-scale irrigation systems is of broad relevance (D. Korten, 1980).

<u>Policy design and organizational design</u>. With the exception of Uma Lele's <u>The Design of Rural Develop-ment: Lessons from Africa</u>, the issues of policy design and the questions concerning organizational design have been treated by different specialists. The communica-tion between those groups has been limited. This is unfortunate because the sets of issues are closely related.

In our discussion of planning and policy analysis, we stressed the advantages of price and market mechanisms as an efficient means of transmitting information and harmonizing the decentralized decisions of large numbers of producers and consumers. We further noted that when markets are reasonably competitive, private firms can perform many of the economic functions satisfactorily with only limited need for government interventions. A major conclusion of that discussion was that government policies and programs should be concentrated on essential activities that are not likely to be undertaken--at least, on the scale that is socially desirable--without government intervention. This suggests that government efforts should be focussed on strengthening the capacity and performance of facilitator organizations such as those for agricultural research and extension that provide "public goods" which permit widespread increases in productivity and output. This also applies, of course, to government organizations providing social services, especially the education, health, nutrition, and family planning programs that we discussed briefly in chapter III.

We have noted the significant advantage of individual family farms as the basic organizational unit for agricultural production. More difficult problems arise in connection with the choice of organizations for performing "commercial functions" such as the marketing of agricultural products, the distribution of farm inputs, and the supply of agricultural credit. Cooperatives or parastatals have frequently been the preferred types of organizations for performing those functions, as we noted earlier. This preference is often justified in terms of protecting small farmers from exploitation by private firms. Additional arguments are advanced in terms of potential gains from economies of scale and from vertical integration of marketing and processing. India's National Dairy Development Board is an interesting example of an organization which has been successful in benefiting small-scale milk producers by realizing those potential gains. Before turning to some of the lessons from that important success story, we review some of the frequent shortcomings of cooperative marketing organizations and the reasons why reliance on private firms may often be a better option.

By and large, cooperatives in developing countries have not fulfilled the hopes that were placed in them. They have frequently been "captured" by the wealthier and more influential members who have obtained a disproportionate share of the resources distributed. This has probably been especially true of credit cooperatives distributing subsidized loans at low interest rates, but also holds for subsidized distribution of farm inputs. In some other cases cooperatives have been essentially instruments of government. This has been common in situations in which the members have little experience or competence in handling the accounting and other tasks involved in gaining a competitive advantage in performing their marketing functions. Some studies of traditional marketing systems have found that intermarket and intertemporal price differences are in line with the costs of transportation and storage, including the cost of tying up capital in stocks in countries where capital is scarce and costly. Excessive price differentials are often due to imperfections in the regional or national marketing system that farmer cooperatives are not likely to be able to resolve (Fox, 1979, p. 312). Because the advantages that cooperatives are able to offer their members are often quite minimal, they are likely to have difficulty in attracting and holding members. When governments are strongly committed to the cooperative approach, they may respond to such problems by giving cooperatives monopoly power as the sole authorized purchasers of major farm products. That type of arrangement may work reasonably well in the marketing of an important export crop; but the outcome is usually unsatisfactory in the case of food crops for the domestic market.

In considering the choice of organizations for handling agricultural marketing, it is well to recognize that there are some practical important arguments for utilizing the organizational capacity and technical and entrepreneurial skills that are often available in private firms. Even in late-developing countries, competition among these private firms frequently limits the monopoly power that they are able to exercise. For example, Lele (1975, p. 114) reports that "most price exploitation that has been observed in Ethiopian markets is covert, through false weights and measures, rather than overt," and this implies that farmers "have the potential to enjoy real bargaining power." These considerations suggest that the most appropriate government action is to facilitate more effective performance of a private or mixed marketing system and to reduce the scope for exploitation by introducing standard weights and measures, by disseminating adequate and reliable price information, and by improving communication and transportation networks.

The shortcomings of quasi-governmental organizations in the marketing of agricultural products tend to be more serious than for cooperatives. There is greater reluctance to permit a parastatal to fail. Hence, governments are likely to subsidize their operations and to give them monopoly power. Two arguments for emphasizing facilitative and regulatory actions by government rather than attempting to replace private firms with a quasi-governmental organization are especially cogent. First, government organizations

such as grain marketing boards tend to be relatively
inefficient, in part because of the need for bureau-
cratic regulations to limit graft and corruption.
Second, the lack of flexibility of such organizations
makes it difficult for them to act in a timely fashion
which is often important in marketing farm products or
distributing inputs: fertilizer that arrives too late
loses a great deal of its value. There is also the
more general consideration that when administrative
manpower is in short supply, government programs should
be concentrated on the higher priority activities such
as making available essential "public goods."

Although it is difficult to create effective
cooperatives, it is important to recognize that
well-managed cooperatives can perform many useful
functions. The system of dairy cooperatives promoted
by India's National Dairy Development Board (NDDB) is
an interesting example. It illustrates some important
lessons about the organization and management of
successful cooperatives--and other organizations as
well.

As we noted in the preceding section, the NDDB has
been highly effective in organizing several thousand
village cooperatives which have enabled over 2 million
small farmers and even landless households to
supplement their incomes by maintaining one or two
cows. Members of a village-level cooperative society
deposit milk twice each day at a collection point
maintained by the society. Special trucks collect the
milk and deliver it to processing centers operated by a
cooperative union that serves some 80 village
societies. The processed dairy products are then sold
in urban centers through a federation of dairy unions.

A significant feature of the management of the
cooperatives included in the NDDB scheme emphasized by
Korten is that "A combination of strong, externally-
audited management systems, daily payments to members,
and public transactions (including tests for quality of
milk) leaves little room for dishonesty on the part of
co-op officials. With little opportunity for corruption, only the more responsible individuals are
attracted to leadership positions." The functions of
the village cooperative society are simple and place
few demands on their leaders and members. Significantly, there are no demands for "communal labor or for
complex decisions that might favor one group over
another." (Korten, 1980, p. 485).

The success of the system of cooperatives promoted
by the NDDB cannot be understood, however, without
considering the evolution of the organization. Its
history traces back to the mid-1940s. At that time a
group of farmers in Gujerat State grew resentful of the
low prices being paid by a private dairy which had a

contract from India's colonial government to purchase and process milk for sale in Bombay. This led to the organization of the Anand Milk Producers Union Limited which, by 1947, had eight village cooperatives with 432 members. In 1949, Verhese Korten was employed in a junior position in the Indian Research Creamery in Anand to advise the Anand union on the purchase of dairy machinery. Later he was assigned to help install the machinery and to train the local staff. Korten then became manager of the cooperative and learned with the farmers how to deal with the organizational and management problems that had to be resolved in order to create an effective prototype program.

A new and larger organization was then developed around the prototype: "appropriate management systems were worked out through experience to meet the demands of the program. The values of integrity, service, and commitment to the poorest member-producers were deeply embedded in its emerging structures. Management staff were hired fresh from school, trained through experience on the job, indoctrinated in the values of the program, and advanced rapidly as it grew" (Korten, 1980, p. 486). This process of bottom-up learning and growth continued for ten years within Gujerat State. When the NDDB was established in 1965, Korten was in charge of the effort to extend the program nationally. He and his coworkers were able to draw upon and apply the lessons and managerial procedures which had been evolved in the process of creating effective village societies and the original dairy union in Anand.

This example has been described in some detail because it suggests that the creation of a successful organization involves a <u>learning process</u> rather than merely applying a blueprint. In fact, as Korten emphasizes, this type of learning process involves three different stages: learning to be <u>effective</u>, learning to be <u>efficient</u>, and learning to <u>expand</u>. Moreover, the learning requires effective interaction between the organization responsible for the program and the beneficiaries; the essential requirement is to achieve a satisfactory "fit" between the organization, the task requirements of the program being implemented, and the needs of the intended beneficiaries (Korten, 1980, pp. 495-501). The experience of the NDDB also illustrates the importance of the competence, dedication, and continuity of leadership in achieving effective performance by development-oriented organizations.

110

NOTES

1/ There is a huge literature dealing with both the theoretical and empirical issues discussed in this and the following paragraphs. Berry and Cline (1979) provide a good summary of both.

2/ Binswanger (1978, pp. 30-42) summarizes a large body of evidence supporting the view that tractor mechanization has very little positive effect on crop yields.

3/ An article and book by Steven Cheung (1968, 1969) set off the lively debate on these issues. A paper by Newberry (1975) is perhaps the most satisfactory analysis. However, many of the main arguments were set forth much earlier by D. G. Johnson (1950).

4/ The statement is by A. K. Sen, one of the outstanding economic theorists who has been concerned with development problems (quoted in Hunter, 1978, p. 37).

5/ To cite one example, a simple projection model for Pakistan considerably underestimated the rate of diffusion of high-yielding varieties but overestimated the rate of increase in yields (Johnston and Cownie, 1969).

6/ A CGIAR review committee report provides a useful summary account of the system (CGIAR, 1981).

7/ The wheeled tool-carrier used at ICRISAT is, however, sophisticated and expensive and therefore beyond the reach of the majority of small farmers unless its cost is borne by several farmers through sharing its use (such as by contract farming). A simpler and much cheaper tool bar now being promoted in Kenya by an FAO/UNDP Agricultural Equipment Improvement Project seems to be more promising for small farmers with limited cash income.

8/ Important contributions to this literature include publications of the Rural Development Committee at Cornell (e.g., Uphoff and Esman, 1974) and the work of D. Korten (1979, 1980, 1981), Chambers (1974), Leonard (1977), Hunter (1978a, 1978b), and Lele (1975). The paragraphs that follow draw very heavily on an excellent treatment of these issues by William C. Clark (Johnston and Clark, in press, chapter 5).

V
Interrelationships Among the Functional Areas and the Determination of Priorities

In chapters II and III, the agricultural development experience of countries characterized by DSS, USF, and "mixed characteristics" models were examined. Two major conclusions are suggested.

First, a broadly based agricultural development involving a large and growing percentage of a country's farm population in the process of technological change is much more effective in expanding employment opportunities and generating increased income and growth of effective demand than a dualistic pattern of development where increases in productivity and output are confined to a small subset of large-scale farm enterprises. Development under the USF model tends to encourage the growth of ever more differentiated factor and product markets. It is the increase in backward, forward, and final demand linkages under this model that tends to propel the development process and expand employment opportunities, output, and effective demand both within and outside the agricultural sector.

In contrast, development under the DSS model is characterized by the use of drastically different technologies by large farmers than those used by the great majority of small farmers. Under these conditions, the growth and development of backward and forward linkages tend to be with a small modern sector that in many cases has stronger linkages to international markets for high technology producer goods and luxury consumption items. Linkages to the domestic manufacturing sector are not only weaker, they are also concentrated in large-scale, capital-intensive firms of a modern sector enclave. As a result, the growth and increasing differentiation of factor and product markets linked to the great mass of small farmers is retarded. In turn, this limits the rate of growth in employment opportunities, output, and effective demand among the poor.

A second major conclusion of the earlier analysis is that the USF and DSS patterns tend to be mutually

111

exclusive. That is, there are serious obstacles in
moving from a DSS to a USF model. Political opposition
to the policies needed to foster a USF pattern is
important. Numerous economic interests seek to
maintain the status quo by impeding the design and
implementation of dispersal strategies essential to the
successful implementation of a USF pattern. The rural
power structure in particular often sees its interests
protected by maintaining the status quo. And, it feels
threatened by attempts to increase the economic
viability of small farms.

Development assistance agencies, which are
independent of the less developed country power
structure, can play an important role in supporting
development strategies that encourage employment and
effective demand generation and discourage strategies
that detract from achieving those objectives. The
logical first step in achieving a USF pattern would be
to undertake land reform. But, this is not a feasible
option in most less developed countries and other
measures are needed. There are other measures that can
be undertaken in the short term that significantly
contribute to the long-term objectives of increasing
employment, output, and effective demand of the rural
poor.

Priority Development Assistance Activities within
Individual Functional Areas

Our analysis of the seven functional areas in
chapter IV assessed the impact of policies and programs
in the various functional areas on the expansion of
employment opportunities and the growth of effective
demand. Here, we suggest some of the most strategic
investment programs and technical assistance that
development assistance agencies might support in each
area. Some activities merit a high priority because
they facilitate progress toward the USF model.

Asset Distribution and Access. A redistributive
land reform is an especially desirable measure for
promoting a USF pattern. Because of strong political
opposition, there will be few opportunities for a major
land reform program. Making a realistic assessment of
the likelihood that a particular country will undertake
a redistributive land reform is a delicate business.
There may be much official rhetoric endorsing land
reform, but governments are often unwilling to incur
the political costs. Political climates, however,
often change rather abruptly. Zimbabwe is an example
where a land redistribution, at least of a modest
nature, is feasible as well as desirable. Support from
different aid-providing agencies and countries in
Zimbabwe is an excellent example of efforts to

facilitate a national goal of land distribution. There may be many such opportunities for the aid donors for a concerted effort to develop a support coalition for a country genuinely committed. Aid-providing agencies may provide the purchase price and or expenses involved in cadastral survey, title transfer, and other elements of a land transfer program.

Development assistance agencies should be cautious in supporting a tenancy reform aimed at setting a legal ceiling on rental payments as a substitute for a redistributive land reform. Because of the great difficulty of effectively enforcing a tenancy reform in the presence of landownership concentration, it is likely to do more harm than good, especially if it encourages landlords to evict their tenants. However, an arrangement for long-term lease of land, say for 15 to 20 years, may be feasible and a viable alternative where the alternatives open to the landlords is political upheaval leading to uncompensated loss of land. A program to encourage or require landlords to share costs of purchased inputs is also likely to be feasible. In a traditional agriculture where use of purchased inputs is of slight importance, such sharing in the cost of nonland inputs is uncommon. However, in situations where farmers are beginning to use high-yield, fertilizer-responsive varieties, there are important gains to be realized from introducing this practice as rapidly as possible. And, the feasibility of cost sharing is enhanced because landlords as well as tenants benefit from increasing output by the use of economically optimal levels of inputs.

It may also be desirable and feasible to support a limited land reform providing landless laborers with "houseplots" large enough to support a kitchen garden and a cow or buffalo. Such a program is likely to be especially useful when accompanied by distribution of nonland assets: poultry, pigs, a cow or two, or hand tools. Use of development assistance funds to support part of the cost of distributing such nonland assets might provide useful leverage in encouraging a government to undertake the modest redistribution of land required to make available houseplots.

Planning and Policy Analysis. It is much easier to assert the need for good planning and policy analysis than to offer constructive suggestions for achieving that capability. The difficulty stems in part from the lack of agreement as to what constitutes good planning and policy analysis. If our critical assessment in chapter IV of the shortcomings of the conventional approach to comprehensive economic planning is valid, there is an urgent need to evolve more effective approaches to planning and policy analysis.

Development assistance agencies can and should play an important role in promoting state of the art planning and policy analysis, especially as it relates to rural development. Priority should be given to evolving an approach along the lines pioneered by W. Granger Morgan, Charles Lindblom, Aaron Wildavsky, and other able practitioners of the art and craft of policy analysis. The need to supplement intellectual cogitation with the techniques of social interactions, as it relates to the development of policies and plans for providing public services, is particularly important. The importance of interactive learning from the analysis and interpretation of past experience and from the feedback that can be derived from strategic monitoring of ongoing programs cannot be underestimated.

AID has long supported research on the organization of the development process and recognized that problems of learning how to implement development programs are as important as determining what to do. An especially important aspect of this interaction between what to do and how to do it involves the decentralization of planning in developing countries. There is a special need to strengthen local planning capabilities and the upward flow of information so that national policies and programs conform more closely to rural needs. Public programs for research, extension, and the construction of infrastructures as well as education, health, and family planning are essential to the development of a small-farm agriculture.

Development and Diffusion of New Technology. A USF pattern of development requires agricultural research and extension programs to find and disperse locally adapted strategies. Development aid donors have already made major contributions to strengthening national programs of agricultural research and extension and to the development of the CGIAR system of international agricultural research centers. Many problems remain.

Four major problem areas limiting development and diffusion of new technology are: (1) the "yield gap" problem related to the large difference in yields obtained by the great majority of farmers and the yields obtained on agricultural experiment stations; (2) underinvestment in national agricultural research programs, especially in tropical Africa; (3) inadequate attention to the special problems of increasing productivity of small farmers under rainfed conditions; and (4) policies that negate positive effects of technological progress on employment opportunities and effective demand.

Some of the most desirable and feasible development assistance investments and technical assistance programs are related to overcoming problems in those four areas. The United States and other industrialized countries have a wealth of experience and expertise in agricultural research and extension; but, there is little scope for direct transfer of technologies evolved in these countries to developing countries. Such technologies are much too capital-intensive to be relevant to small-scale farmers in less developed countries. A major strength of the international agricultural research centers is their orientation toward applied research. Scientists at the international centers can draw upon advanced scientific knowledge and research techniques evolved in the United States and other developed countries while concentrating their own research on biological, chemical, and related innovations that are divisible and potentially suitable for small farmers. In most cases, however, strong national and regional research programs must carry out the adaptive research needed to develop improved varieties and agronomic practices suited to a variety of agroclimatic environments. There is a continuing need for economic and technical assistance for strengthening those national research programs.

More effective techniques are needed to ensure that the technological innovations generated by research are feasible and profitable given the conditions and constraints faced by small farmers. Adaptive onfarm research on farming systems, mentioned in chapter IV, is an example of a methodology that may achieve this objective. There is also a high-priority need to increase effectiveness of agricultural extension programs in transferring useful knowledge to farmers. The Training and Visit system pioneered by Daniel Benor and the World Bank seems to be a promising approach.

The need for more effective research methodologies is especially critical in rainfed areas where the yield potential of improved seed-fertilizer combinations are much more limited than under irrigated conditions. Research at ICRISAT and elsewhere suggests that it is often necessary to introduce farm equipment and tillage innovations to improve soil and water management in order to realize the yield potential of higher yielding varieties and fertilizers.

An FAO/UNDP Agricultural Equipment Improvement Project in Kenya is one of the few serious attempts to evolve an effective methodology for R&D aimed at simple, inexpensive farm implements and wider, more effective use of animal draft power. These R&D activities have been carried out in several ecological regions, identifying implements fitting different types of farming systems.

Kenya's Ministry of Agriculture is training small farmers in the use of the most promising items of equipment at selected farmer training centers. An AID

grant has enabled the Ministry to place orders for the
local manufacture of those implements for use in
demonstration and training programs and also for sale
at cost to farmers. One criterion used in the
evaluation process is the suitability of the equipment
for local production. It is hoped that demand by
farmers will stimulate growth of local manufacture of
farm implements. This will ideally initiate an interactive process leading to gradual increases in farm
productivity and in the manufacturing capabilities of
small and medium-scale workshops. We believe that aid
providing agencies should give a high priority to
monitoring this Kenyan experience to verify this R&D
methodology.

A final recommendation is that development
assistance agencies should be extremely cautious about
supporting imports of tractors and other capital-
intensive, labor-displacing technologies. This caution
is clearly most important in the case of the lower
income countries where agriculture's share of the labor
force is still large and the economically active
population in the agricultural sector is still
increasing.

Investment in Rural Infrastructure. There have been
conflicting views in the aid-providing countries and
agencies about support for the construction of rural
roads, irrigation, and other types of infrastructure.
Some, for example, have argued against assistance for
such projects because they do not confer direct benefits on the rural poor. A more appropriate response to
the concern about problems of rural poverty has been
the emphasis in some of the more recent infrastructure
projects on the use of labor-intensive construction
methods supported by AID, the World Bank, and ILO.
This approach seems to have been particularly successful in projects for building rural access roads.

The highway departments in less developed countries
have often been reluctant to adopt labor-intensive
techniques, and AID and World Bank funding procedures
have also tended to encourage equipment-based
projects. It has been found possible, however, to
modify organizational and management techniques in
these countries in order to facilitate the use of
labor-intensive construction methods. Priority should
continue to be given to such projects rather than
projects employing capital-intensive techniques and
extensive use of road-building equipment.

Further expansion of irrigation and drainage
facilities is critically important for the expansion of
agricultural production and employment, especially in
Asia. External support must be made available for
well-designed projects; funding may come from the World
Bank, the Asian Development Bank, and AID. There may

be attractive opportunities for aid donor countries to influence the design and management of irrigation projects so that they will have a maximum impact on productivity and employment. Priority should be given to rural electrification when availability of electric power will foster a substantial increase in rural, nonfarm employment. Rural works projects, discussed later as one of the more promising ways to use food aid to meet short-term needs, also contribute to longer-term strategic objectives.

Marketing and Storage, Input Supply, and Credit. Expansion of production by small farmers clearly depends on satisfactory performance of these support functions. We believe, however, that the essentially commercial functions of marketing farm products and distributing inputs do not merit priority in the allocation of government resources in many cases. Support should be concentrated on assistance for strengthening facilitative and regulatory actions, such as improving facilities for public markets and storage, introducing standard weights and measures, and disseminating reliable price information.

Provision of credit to small farmers calls for more direct government action and warrants a relatively high priority. A flexible approach is called for in determining the institutional mechanisms--government, cooperative, informal groups of farmers, or banks--that will be most effective in channeling credit to small farmers. Development assistance programs should avoid artificially low interest rates creating the "excess demand" that requires administrative rationing. The larger and more powerful farmers will receive most of the subsidized credit under such conditions.

Rural Industry and Ancillary Activities. The certainty that the rural population of working age will continue to increase underscores the importance of expanding nonfarm employment opportunities. Even a USF pattern is incapable of expanding productive employment in agriculture rapidly enough to absorb the growing rural labor force, especially when the number of landless laborers is already large. Stimulating the growth of rural-based industries is also important because expanded output of appropriate farm equipment and consumer goods can make a notable contribution to raising farm productivity and strengthening producer incentives.

The most important requirement for stimulating growth of rural-based industries is generating wide-spread increases in the income and effective demand of the farm population. There are, however, other ways to foster rural industrialization, such as training and extension programs to upgrade technical skills of small and medium-scale workshops. R&D activities such as the

Kenya program discussed earlier to develop and diffuse simple items of farm equipment are also highly relevant. An important opportunity lies in encouraging reforms in a country's macroeconomic policies which obstruct creation and growth of small, labor-intensive firms while directly or indirectly subsidizing large-scale, capital-intensive firms.

Development assistance agencies can also support various ancillary activities carried out by members of farm households. These activities can quickly increase employment, income, and effective demand among small farmers and even landless families if they have the "houseplots" discussed earlier. Dairy, poultry, pigs, fish ponds, woodlots and charcoal, and handicrafts are possibilities. The principal role of aid agencies in supporting ancillary activities is likely to be in assisting the subsidized distribution of nonland assets such as a cow or baby chicks or pigs.

Institutional Development: Improving Organizational Structures and Managerial Procedures. Strengthening local participation and the performance of various "facilitator organizations" is critical to providing the public goods such as research, extension, and irrigation systems.

Improving administrative capabilities in a developing country is inevitably a difficult, time-consuming process. There is an urgent need to build the competence of research, extension, health, and other facilitator organizations and also to evolve effective methods of fostering local organizations. Much discussion of "participation" has remained at the level of rhetoric. There has been an unfortunate tendency to regard participation as a free good and to ignore the fact that it requires a considerable investment of time and energy on the part of participants. Efforts to support local organizations and greater decentralization in development planning and implementation will be more effective if serious attention is given to the factors important in inducing people to invest their time and energy in group activities. Group or collective farming, advocated in a number of less developed countries, creates extremely difficult organizational problems because of the complexity of determining individual responsibilities, distributing benefits, and avoiding poor work performance. Consequently, the perceived benefits from such an organizational effort are unlikely to exceed costs. On the other hand, creating or strengthening local organizations to participate in planning and managing an irrigation system is likely to yield substantial benefits.

Use of Food Aid to Achieve Long-term Development Objectives 2/

The main objective of food aid is to provide long-term development for the developing countries; but, such aid plays different roles in different countries. For example, the USF-pattern countries place broadly based agriculture in a leading position providing directly or generating indirectly food production, employment, and income growth. Food aid in such countries helps to expand productive employment, thereby increasing the effective demand for food and improving food consumption without generating self-defeating inflation.

If a country already pursues a relatively equity-oriented development strategy or tries to follow the USF pattern, food aid can be used to support complementary, nonagricultural activities such as a public works program. The public works programs are often known as "work for food" programs because payments of wages are normally made in kind, using grain checks with the governments.

Food aid can also be used to support policy changes designed to promote land tenure reform and trade liberalization. For example, during the land redistribution, food aid can serve as short-term security to the farmers.

Food aid can support accelerated agricultural development and increased food production by: (1) permitting more employment of poor people (thereby increasing effective demand for food) in a non-inflationary environment; (2) stabilizing agricultural prices and incomes, thereby reducing price uncertainties and stimulating increased investment in agriculture; (3) financing physical infrastructure and related services needed to complement increased agricultural production; (4) financing construction of storage facilities; and (5) inducing land reform, creating permanent employment-intensive activities, and, most important, facilitating more equitable food distribution.

The Role of the Private Sector

Government policies and intervention programs should concentrate on those activities most important to the development process and not likely to be undertaken without government action. This dictum applies to national economic policies related to trade, land tenure, prices, and taxation. And, it applies to a variety of "public goods" ranging from agricultural research and extension to public health and family planning programs.

Price and market mechanisms in transmitting
information and in harmonizing the decisions of
millions of farmers and many other producers of goods
and services are more appropriate for most production
processes and for commercial functions such as
marketing agricultural products and distributing farm
inputs. Decentralized decisionmaking based on intimate
knowledge of local conditions has great advantages in
securing efficient use of scarce resources to maximize
output and minimize costs. Provided that markets are
reasonably competitive, private firms have significant
advantages in satisfying society's needs. Profit
maximization provides a strong and clearcut incentive
to minimize costs of production and also permits the
flexibility required to make quick decisions in
response to changing conditions. Reliance on
individual family farms and other private production
units has the great advantage of economizing on scarce
administrative talent.

Political leaders and policymakers in many less
developed countries reject this favorable view of the
role of prices and markets. Some of the most serious
failures of a market system in many developing
countries result from the common practice of "rigging"
markets so that prices, especially of capital and
foreign exchange, are distorted to favor the privileged
modern sector firms that have access to those resources
at artificially low prices. When the distribution of
wealth, including investment in human capital, is
highly unequal, a market system will perform badly when
judged by the criterion of equity. Efforts to correct
that situation by relying on government-administered
prices are, however, likely to compound and perpetuate
inequality as well as lead to inefficiencies in the use
of scarce resources. Such lessons recommend a USF
pattern which fosters more rapid growth of employment
opportunities and more widespread and equal growth of
incomes and effective demand. Also important are
programs providing wider access to education and health
services and thereby leading to a more equitable
distribution of human capital.

Attempts by governments to undertake tasks better
performed by private firms will make it more difficult
for government organizations to fulfill adequately
those essential functions which will not be performed
adequately, if at all by private firms.

The foregoing arguments do not apply with as great
a force to large transnational corporations as they do
to the numerous domestic firms. The economic
advantages of large transnational corporations derive
primarily from their economies of scale in preforming
selected economic functions. Those scale economies are
not limited to the production process. They also result

from the greater ability of such firms to invest in R&D, their market connections and greater experience in advertising and other techniques of market promotion, and their access to capital on relatively favorable terms in international capital markets. In the case of technologies that are inherently large-scale and capital-intensive, such as the manufacture of nitrogen fertilizers, transnational corporations can play a very useful role in the design and manufacture of efficient, low-cost factories. However, less developed countries would be better off to import nitrogen fertilizer rather than tie up a large amount of capital in a fertilizer factory unless the principal raw material natural gas is locally available and cheap. For certain export commodities, especially a perishable commodity such as bananas, transnational corporations have significant advantages in organizing the assembly, packing, transporting, marketing, and quality controlling essential to exporting.

Against those potential advantages, however, large foreign corporations often transfer technologies and products inappropriate to conditions in less developed countries. In fact, many of the large-scale firms in a developing country's "modern-sector" enclave are foreign corporations that benefit from tax holidays, allocations of foreign exchange at an overvalued official exchange rate, and other privileged treatment. In contrast, a major benefit of decentralized growth of small- and medium-scale domestic firms is that they are under greater pressure to adopt labor-intensive technologies which minimize on requirements for capital and foreign exchange and therefore generate more rapid expansion of job opportunities for the large and growing domestic labor force.

Efforts by aid donors to influence less developed country policies on the role of the private sector will require a great deal of sensitivity. A doctrinaire approach and attempts to lecture LDC leaders concerning the virtues of free enterprise are likely to be counter productive. Questions related to the respective roles of government and private firms are politically sensitive issues in any country. Promoting research and good policy analysis that focuses on assessing the performance of public, cooperative, and private organizations can promote a more pragmatic view of the strengths and weaknesses of the different types of organizations in performing various types of functions. This can encourage a greater acceptance of the need for a pluralistic approach to issues of organizational design.

VI
Summary

Expanded employment opportunities and increased effective demand of the rural poor for food and other essential commodities are of critical importance in less developed countries. But, unemployment, underemployment, rural poverty, and rapid population growth pose especially formidable problems for governments in the lower income developing countries where 60 to 80 percent of the population and labor force still depend on agriculture. This concentration of poverty in rural areas will likely persist for two, three, or more decades. Under those conditions, efforts to rely on shortrun, tactical solutions are doomed.

A long-term strategic approach is needed. Central to our proposed long-term strategic approach is an emphasis on fostering broadly based agricultural development strategies that approximate as closely as possible the USF model of agricultural development. Success in achieving a USF pattern requires an emphasis on dispersal strategies that will foster widespread increases in productivity among a large and growing percentage of the small farm units. This, in turn, necessitates an emphasis on labor-using, capital-saving technologies which facilitate rapid increases in opportunities for productive employment both within and outside agriculture.

A USF pattern is difficult to achieve for a number of interrelated reasons. Powerful political groups have a vested interest in the economic dualism characterizing the alternative DSS pattern of development. Moreover, there is only limited recognition of the extent to which the USF and DSS models are mutually exclusive alternatives, given the structural and demographic characteristics of lower-income developing countries.

Formidable technological and organizational problems must be overcome in order to design and implement dispersal strategies. Japan and Taiwan, outstanding examples of successful USF patterns, were able to rely mainly on the divisible innovations represented by high-yielding varieties and increased use of fertilizers under the relatively homogeneous and controlled conditions of irrigated production of rice and a few other major crops. In contrast, farming in a great many developing countries is carried out under rainfed conditions. Dispersal strategies must, therefore, be developed for a variety of agroclimatic conditions. And, there is a strong presumption that biological chemical innovations will have to be associated with equipment and tillage innovations to improve soil and water management. Effective organizations and a capacity to tap local knowledge of farming systems and constraints faced by farmers are more demanding because of the need to implement a variety of location-specific dispersal strategies.

Success or failure in implementing a USF pattern will depend on effective but selective attention to activities in seven functional areas: asset distribution and access; planning and policy analysis; development and diffusion of new technology; rural infrastructure; marketing and storage, input supply, and credit; rural industry and ancillary activities; and improving organizational structures and managerial procedures (institutional development). Because of the acute scarcity of money and trained manpower and insufficient administrative capacity, there are bound to be competitive tradeoffs severely limiting the range of activities feasible at any one point. Hence, aid-recipient governments and aid-giving agencies must make hard choices about the allocation of available resources. The difficulty of making optimal choices is aggravated, however, by the fact that it is equally important to take account of significant complementarities among the functional areas and certain activities within individual areas, such as the concurrent need for improved seed-fertilizer combinations and for equipment and tillage innovations. The importance of slowing the rate of growth of population and thus the labor force points to the priority need for the interrelated goals of health and family planning.

Development assistance agencies must take a long-term, strategic view in making judgments about development assistance priorities. Foreign aid programs are subject to political constraints; a long-term strategic plan must be developed for each country program that realistically accounts for these constraints. Also, development assistance activities can be designed and initiated in the short term that will contribute to achieving long-term objectives. But, to be most effective, these activities should be undertaken within the context of a long-term plan designed to achieve eventually a USF development pattern. Assistance providers should adopt a policy position that ensures that food aid is associated with economic and technical assistance designed to expand domestic food production and encourage a USF pattern of development.

NOTES

1/ William C. Clark, one of the ablest of the younger generation of policy analysts, has made an important initial effort in that direction in his chapter on "Policy Analysis and the Development Process" (Johnston and Clark, 1982, chapter 1).

2/ For a detailed discussion of this subject see "Food Aid and Development," 1981, Washington, DC PPC/AID.

References Cited

D. W. Adams, "The Economics of Land Reform: Comment,"
Food Research Institute, 12, 2, 1973. Agency for
International Development, Agricultural Development
Policy Paper. Washington, D.C.: AID, 1978.

Agency for International Development, Agricultural
Development Policy Paper. Washington, D.C.: AID,
1978.

_____ Food AID and Development, 1981.

AID/San Jose, Programa de Desarrollo Agropecuaria
1971-1974. San Jose, Costa Rica: Agencia para el
Desarrollo Internacional, Diciembre de 1970.

M. S. Ahluwalia and H. Chenery, "The Economic
Framework". In Redistribution with Growth, by H.
Chenery et al., London: Oxford University Press,
1974.

K. R. M. Anthony, B. F. Johnston, W. O. Jones, and V.
C. Uchendu, Agricultural Change in Tropical Africa.
Ithaca, N.Y.: Cornell University Press, 1979.

K. J. Arrow, The Limits of Organization. New York:
Norton, 1974.

B. U. Bagadion and F. F. Korten, "Developing Viable
Irrigators' Associations: Lessons from Small Scale
Irrigation Development in the Philippines",
Agricultural Administration, 7, 4, November 1980.

P. Bardhan and A. Rudra, "Terms and Conditions of
Sharecropping Contracts: An Analysis of Village
Survey Data in India", Journal of Development
Studies, 16, 3, April 1980.

C. Barlow, The National Rubber Industry: Its
Development, Technology, and Economy in Malaysia.
Oxford: Oxford University Press, 1978.

_____ and B. F. Johnston, "Converging Views on
Strategies for Agricultural Development: The
Focus/Bimodal and Dispersal/Unimodal Alternative",
draft manuscript, July 1981.

R. H. Bates, "States and Political Intervention in Markets: A Case Study from Africa". Social Science Working Paper 345. Pasadena: California Institute of Technology, Division of the Humanities and Social Sciences, September 1980.

J. M. Beeney, UNDP Report to the United Republic of Tanzania on Agricultural Mechanization. Rome: FAO, 1975.

C. L. G. Bell and P. B. R. Hazell, "Measuring the Indirect Effects of an Agricultural Investment Project on Its Surrounding Region", American Journal of Agricultural Economics, 62, 1, February 1980.

C. Bell, P. Hazell, and R. Slade, "The Prospects for Growth and Change in the Muda Region". In The Political Economy of Rice and Water: Village-Level Modernization, Employment and Income Distribution in Southeast Asia, ed., G. B. Hainsworth, forthcoming.

D. Benor and J. Q. Harrison, Agricultural Extension: The Training and Visit System. Washington, D.C.: World Bank, May 1977.

R. A. Berry and W. R. Cline, Agrarian Structure and Productivity in Developing Countries. Baltimore: Johns Hopkins University Press, 1979.

H. P. Binswanger, The Economics of Tractors in South Asia: An Analytical Review. New York: Agricultural Development Council; Hyderabad, India: International Crops Research Institute for the Semi-Arid Tropics, 1978.

H. P. Binswanger, R. D. Ghodake, and G. E. Thierstein, "Observations on the Economics of Tractors, Bullocks, and Wheeled Tool Carriers in the Semi-Arid Tropics of India". In Proceedings of the International Workshop on Socioeconomic Constraints to Development of Semi-Arid Tropical Agriculture, 19-23 February 1979, Hyderabad, India, by ICRISAT (International Crops Research Institute for the Semi-Arid Tropics). Patancheru, A.P. India: ICRISAT, 1980.

H. P. Binswanger and M. R. Rosenzweig, Contractual Arrangements, Employment and Wages in Rural Labor Markets: A Critical Review. New York: Agricultural Development Council, Inc., 1981.

H. I. Blutstein, L. C. Andersen, E. C. Betters, J. H. Dombrowski, and C. Townsend, Area Handbook for Costa Rica. Washington, D.C.: U.S. Government Printing Office, October 1970.

A. A. Bodenstedt et al., Agricultural Mechanization and Employment. Heidelberg: Research Centre for International Agrarian Development, 1977.

J. K. Boyce and R. E. Evenson, National and International Agricultural Research and Extension Programs. New York: Agricultural Development Council, 1975.

J. M. Brewster, "The Machine Process in Agriculture and Industry", Journal of Farm Economics, 32, 1, February 1950.

R. Chambers, Managing Rural Development: Ideas and Experience From East Africa. Uppsala: Scandinavian Institute of African Studies, 1974.

H. B. Chenery, "Poverty and Progress: Choices for the Developing World". In Poverty and Basic Needs. Washington, D.C.: World Bank, September 1980.

S. N. S. Cheung, "Private Property Rights and Sharecropping", Journal of Political Economy, 76, 6, November/December 1968.

_____ The Theory of Share Tenancy. Chicago: University of Chicago Press, 1969.

China, Republic of, Taiwan Statistical Data Book 1974. Taipei: Economic Planning Council, June 1974.

C. H. Chiu, "Food Supply and Nutrition Requirements in Taiwan". Taipei: JCRR, Rural Health Division, August 20, 1976.

E. Chuta and C. Liedholm, Rural Non-Farm Employment: A Review of the State of the Art. MSU Rural Development Paper No. 4. East Lansing: Michigan State University, Department of Agricultural Economics, 1979.

Consultative Group on International Agricultural Research (CGIAR), Report of the Review Committee. Washington, D.C.: CGIAR, September 1981.

R. A. Dahl and C. E. Lindblom, Politics, Economics, and Welfare. New York: Harper and Brothers, 1953.

J. J. de Veen, The Rural Access Roads Programme: Appropriate Technology in Kenya. Geneva: International Labour Office, 1980.

B. Duff, "Providing Assistance in the Mechanization of Small Farms". Paper prepared for presentation at the Seminar on the Mechanization of Small-Scale Peasant Farming, Sapporo, Japan, July 7-12, 1980.

R. E. Evenson, "The Organization of Research to Improve Crops and Animals in Low Income Countries". In Distortion in Agricultural Incentives, ed., T. W. Schultz. Bloomington and London: Indiana University Press, 1978.

FAO, Assistance in Agricultural Mechanization, Tanzania. Mission Report, September-October, 1974. Rome: FAO, 1975.

_____ The State of Food and Agriculture, 1981. Rome: FAO.

R. Fox, "Potentials and Pitfalls of Product Marketing Through Group Action by Small-Scale Farmers", Agricultural Administration, 6, 4, October 1979.

M. Franda, Small is Politics: Organizational Alternatives in India's Rural Development. New Delhi: Wiley Eastern Limited, 1979.

130

J. K. Galbraith, The Nature of Mass Poverty. Cambridge,
Mass.: Harvard University Press, 1979.
W. Galenson, "The Labor Force, Wages, and Living
Standards". In Economic Growth and Structural Change
in Taiwan: The Postwar Experience of the Republic of
China, ed., W. Galenson. Ithaca and London: Cornell
University Press, 1979.
J. Gerhardt, The Diffusion of Hybrid Maize in Western
Kenya--Abridged by CIMMYT. Mexico City: Centro
Internacional de Mejoramiento de Maize y Trigo, 1975.
R. D. Ghodake, J. G. Ryan, and R. Sarin. Human Labor
Use in Existing and Prospective Technologies of the
Semi-Arid Tropics of Peninsular India. Progress
Report, Economics Program-1, Village Level Studies
series 1.3 Hyderabad, India: International Crops
Research Institute for the Semi-Arid Tropics,
December 1978.
R. H. Goldman, "Staple Food Self-Sufficiency and the
Distributive Impact of Malaysian Rice Policy", Food
Research Institute Studies, 14, 3, 1975.
K. Griffin, The Political Economy of Agrarian Change.
London: Macmillan Press Ltd., 1979.
R. D. Hansen, The Politics of Mexican Development.
Baltimore: Johns Hopkins University Press, 1971.
Y. Hayami and V. W. Ruttan, Agricultural Development:
An International Perspective. Baltimore and London:
Johns Hopkins Press, 1971.
O. Hesselmark, "Appendix II. The 1974 Kenya Maize
Farmers Survey". In The Diffusion of Hybrid Maize in
Western Kenya--Abridged by CIMMYT, by J. Gerhardt.
Mexico City: Centro Internacional de Mejoramiento
de Maiz y Trigo, 1975.
P. E. Hildebrand, Generando Tecnologia para Agricultores
Tradicionales: Una Metodologia Multidisciplinaria
(Generating Technology for Traditional Farmers: A
Multidisciplinary Methodology). Guatemala:
Socioeconomia Rural, Instituto de Ciencia y
Tecnologia Agricolas, Sector Agricola, 1976.
S. C. Hsieh and T. H. Lee, Agricultural Development
and Its Contributions to Economic Growth in Taiwan.
Economic Digest Series No. 17. Taipei: Joint
Commission on Rural Reconstruction, 1966.
G. Hunter, "Report on Administration and Institutions".
In Asian Development Bank, Rural Asia: Challenge and
Opportunity, Supplementary Papers, Vol. IV,
Administration and Institutions in Agricultural and
Rural Development. Manila: Asian Development Bank,
1978.
_____ "Guidelines". In Agricultural Development
and the Rural Poor, ed., G. Hunter. London:
Overseas Development Institute, 1978.

G. Hyden, Beyond Ujamaa in Tanzania: Underdevelopment
and an Uncaptured Peasantry. Berkeley and Los
Angeles: University of California Press, 1980.
India, Government of, Draft Five-Year Plan 1978-83.
New Delhi: Planning Commission, 1978.
Institut de Recherches Agronomiques Tropicales (IRAT),
Les Systemes de Culture du Riz Pluvial. Paris: 1984.
International Bank for Reconstruction and Development,
The Economic Development of Tanganyika. Baltimore:
Johns Hopkins University Press, 1961.
International Labour Office, Towards Self-Reliance:
Development Employment and Equity Issues in
Tanzania. Addis Ababa: ILO Jobs and Skills
Programme for Africa, 1978.
International Maize and Wheat Improvement Center
(CIMMYT), Economics Program, Planning Technologies
Appropriate to Farmers: Concepts and Procedures.
Mexico City: CIMMYT, 1980.
N. S. Jodha, "Effectiveness of Farmers' Adjustments to
Risk", Economic and Political Weekly, 13, 25, June
24, 1978.
D. G. Johnson, "Resource Allocation under Share
Contracts", Journal of Political Economy, 58, 2,
April 1950.
B. F. Johnston, "Agricultural Development and Economic
Transformation: A Comparative Study of the Japanese
Experience", Food Research Institute Studies, 3, 3,
November 1962.
_____ "Changes in Agricultural Productivity".
In Economic Transition in Africa, eds., M. J.
Herskovits and M. Harwitz. Evanston, Illinois:
Northwestern University Press, 1964.
_____ "Farm Equipment Innovations and Rural
Industrialization in Eastern Africa: An Overview".
World Employment Programme Research Working Paper WEP
2-22.WP.80. Geneva: International Labour
Organization, July 1981.
B. F. Johnston and W. C. Clark. Redesigning Rural
Development: A Strategic Perspective. Baltimore:
Johns Hopkins University Press, 1982.
B. F. Johnston and J. Cownie, "The Seed-Fertilizer
Revolution and Labor Force Absorption", American
Economic Review, 59, 4, September 1969.
B. F. Johnston and P. Kilby. Agriculture and Structural
Transformation: Economic Strategies in Late-
Developing Countries. New York: Oxford University
Press, 1975.
R. K. Karanjia. The Mind of Mr. Nehru. London: George
Allen and Unwin, 1960.
Kenya, Republic of. Development Plan for the Period
1979 to 1983. Nairobi: 1979.

P. Kilby. Industrialization in an Open Economy:
Nigeria 1945-1966. Cambridge: Cambridge University
Press, 1969.

P. Kilby, ed., Entrepreneurship and Economic
Development. Glencoe, Illinois: Free Press, 1971.

D. C. Korten, "Toward a Technology for Managing Social
Development". In Population and Social Development
Management: A Challenge for Management Schools, ed.
D.C. Korten. Caracas: Instituto de Estudio
Superiores de Administracion-IESA, 1979.

——— "Community Organization and Rural
Development: A Learning Process Approach", The
Public Administration Review, 40, 5, September/
October 1980.

——— "Social Development: Putting People
First". In Bureaucracy and the Poor: Closing the
Gap, eds., D. C. Korten and F. B. Alfonso.
Singapore: McGraw-Hill, 1981.

R. Krishna, Fiscal Measures for Employment Promotion in
Developing Countries. Geneva: International Labour
Office, 1972.

——— "Measurement of the Direct and Indirect
Employment Effects of Agricultural Growth with
Technical Change". Paper prepared for the Seminar on
Technology of Employment. New Delhi: Ford
Foundation, March 21-24, 1973.

——— "A Framework of Rural Credit Policy for the
Small Farmers of Asia". Keynote address delivered at
the Second General Assembly of Asian and Pacific
Regional Agricultural Credit Association. Karachi,
Pakistan: February 5, 1979.

S. Kuznets. Economic Growth of Nations: Total Output
and Production Structure. Cambridge, Mass.: Harvard
University Press, 1971.

T. H. Lee. Intersectoral Capital Flows in the Economic
Development of Taiwan, 1985-1960. Ithaca, New York:
Cornell University Press, 1971.

U. Lele. The Design of Rural Development: Lessons from
Africa. Baltimore and London: Johns Hopkins
University Press, 1975.

——— "Rural Africa: Modernization, Equity, and
Long-Term Development", Science, 211, 4482, February
6, 1981.

D. K. Leonard. Researching the Peasant Farmer:
Organization Theory and Practice in Kenya. Chicago
and London: University of Chicago Press, 1977.

C. E. Lindblom. Politics and Markets. New York: Basic
Books, 1977.

M. Lipton. Why Poor People Stay Poor: A Study of Urban
Bias in World Development. London: Temple Smith,
1977.

_____ "Inter-Farm, Inter-Regional, and Farm-Non-Farm Income Distribution: The Impact of the New Cereal Varieties", World Development, 6, 3, 1978.

M. F. Lofchie, "Agrarian Crisis and Economic Liberalisation in Tanzania", Journal of Modern African Studies, 16, 3, 1978.

M. K. Lowdermilk, "Diffusion of Dwarf Wheat Production Technology in Pakistan's Punjab". Ph.D. dissertation, Cornell University, 1972.

J. H. J. Maeda, "Creating National Structures for People Centered Agrarian Development". In Bureaucracy and the Poor: Closing the Gap, eds., D. C. Korten and F. B. Alfonso. Singapore: McGraw-Hill, 1981.

L. J. Mata and E. Mohs, "As Seen from National Levels: Developing World". In Progress in Human Nutrition, eds., S. Margen and R. A. Ogar. Vol. 2. Westport, Connecticut: AVI Publishing, 1978.

M. G. Morgan, "Bad Science and Good Policy Analysis", Science, 201, 4350, September 15, 1978.

B. Mukhoti, "Agrarian Structure in Relation to Farm Investment Decisions and Agricultural Productivity in a Low-Income Country: The Indian Case", Journal of Farm Economics, 48: 1210-15, December 1966.

B. Mukhoti, "Agrarian Structure in Relation to Farm Investment Decisions and Agricultural Productivity in a Low Income Country: The Indian Case: Reply", American Journal of Agricultural Economics, Vol. 50, No. 4, November 1968.

B. Mukhoti, "Economic Development and Non-Agricultural Income of the Poor: Some Theoretical Considerations", presented at Allied Social Science Convention, Atlanta, December 1979.

B. Mukhoti, "New Economics of Development: Agrarian Based Strategy for India", Paper prepared for and published in The Papers and Proceedings of the Second Conference of the Association of Indian Economics Studies, Montclair, New Jersey, August 19-21, 1977.

B. Mukhoti, "Patterns of Technological Transformation of Agriculture and Economic Development", presented at Allied Social Science Convention, Denver, Colorado, September 5-7, 1980.

B. Mukhoti, "Some Theoretical Aspects of Disguised Unemployment", Indian Journal of Economics, 59, 233, Part II, October 1978.

D. Narain and S. Roy, Impact of Irrigation and Labor Availability on Multiple Cropping: A Case Study of India. Research Report 20. Washington, D. C.: International Food Policy Research Institute, November 1980.

134

D. Newbery, "The Choice of Rental Contract in Peasant Agriculture". In Agriculture in Development Theory, ed., L. Reynolds. New Haven: Yale University Press, 1975.

J. K. Nyerere. Ujamaa--Essays on Socialism. Dar es Salaam: Oxford University Press, 1968.

K. Ohkawa, B. F. Johnston, and H. Kaneda, eds. Agriculture and Economic Growth: Japan's Experience. Toyko: University of Tokyo Press, 1969.

P. Oram, J. Zapata, G. Alibaruho, and S. Roy, Investment and Input Requirements for Accelerating Food Production in Low-Income Countries by 1990. Research Report 10. Washington, D.C.: International Food Policy Research Institute, September 1979.

D. Perkins et al. Rural Small-Scale Industry in the People's Republic of China. Berkeley: University of California Press, 1977.

S. B. Peterson, "The State and the Organization Infrastructure of the Agrarian Economy". In Linkages to Decentralized Units, by D. Leonard, D. Marshall, J. Garzon, S. Peterson, and Sven Steinmo. Berkeley: University of California, Project on Managing Decentralization, February 1981.

J. E. Potter, M. Ordonez G., and A. R. Meashem, "The Rapid Decline in Colombian Fertility", Population and Development Review, 2, 3-4, September-December 1976.

C. E. Pray, "The Green Revolution as a Case Study in Transfer of Technology", The Annals of the American Academy of Political and Social Science, 458, November 1981.

T. G. Rawski. Economic Growth and Employment in China. New York: Oxford University Press, 1979.

J. G. Ryan, R. Sarin, and M. Pereira, "Assessment of Prospective Soil-, Water-, and Crop-Management Technologies for the Semi-Arid Tropics of Peninsular India". In Proceedings of the International Workshop on Socioeconomic Constraints to Development of Semi-Arid Tropical Agriculture, 19-23 February 1979, Hyderabad, India, by ICRISAT (International Crops Research Institute for the Semi-Arid Tropics). Patancheru, A. P., India: ICRISAT, 1980.

C. J. Saenz, "Population Growth, Economic Progress, and Opportunities on the Land: The Case of Costa Rica". Research Paper No. 47. Madison, Wisconsin: Land Tenure Center, June 1972.

J. Samoff, "Crises and Socialism in Tanzania", Journal of Modern African Studies, 19, 2, June 1981.

G. E. Schuh, "Approaches to 'Basic Needs' and to 'Equity' that Distort Incentives in Agriculture". In Distortions of Agricultural Incentives, ed., T. W. Schultz. Bloomington and London: Indiana University Press, 1978.

W. M. Senga, "Kenya's Agricultural Sector". In Agricultural Development in Kenya, eds., J. Heyer, J. K. Maitha, and W. M. Senga. Nairobi: Oxford University Press, 1976.

M. M. Shah and F. Willekens, Rural-Urban Population Projections for Kenya and Implications for Development. Laxenburg, Austria: International Institute for Applied Systems Analysis, 1978.

H. A. Simon, "Designing Organizations for an Information-Rich World". In Computers, Communications and the Public Interest, ed., M. Greenberger. Baltimore: Johns Hopkins Press, 1971.

T. C. Smith. The Agrarian Origins of Modern Japan. Stanford, California: Stanford University Press, 1959.

E. Stanley and R. Morse. Modern Small Industry for Developing Countries. New York: McGraw-Hill, 1965.

G. Standing, Labour Force Participation and Development. Geneva: International Labour Office, 1978.

M. Tamin, "Rice Self-Sufficiency in West Malaysia: Microeconomic Implications", Ph.D. Dissertation, Stanford University, March 1978.

J. Tendler, Rural Electrification: Linkages and Justifications. AID Program Evaluation Discussion Paper No. 3. Washington, D.C.: Agency for International Development, April 1979.

_____, New Directions Rural Roads. AID Program Evaluation Discussion Paper No. 3. Washington, D.C.: Agency for International Development, March 1979.

E. Thorbecke, "Agricultural Development". In Economic Growth and Structural Change in Taiwan: The Postwar Experience of the Republic of China, ed., W. Galenson. Ithaca and London: Cornell University Press, 1979.

V. C. Uchendu and K. R. M. Anthony, Agricultural Change in Geita District. Nairobi: East African Literature Bureau, 1974.

_____, Agricultural Change in Kisii District: Kenya. Nairobi: East African Literature Bureau, 1975.

United States Department of Agriculture, Task Force report on New Directions for U.S. Food Assistance, prepared for the Committee on Agriculture, Nutrition, and Forestry. United States Senate, 1978.

N. T. Uphoff and M . J. Esman, Local Organization for Rural Development: Analysis of Asian Experience. Special Series on Rural Local Government RLG No. 19. Ithaca, N.Y.: Rural Development Committee, Center for International Studies, Cornell University, 1974.

A. Valdes, "The Transition to Socialism: Observations on the Chilean Agrarian Reform". In Employment in Developing Nations, ed., E. O. Edwards. New York: Colombia University Press, 1974.

V. S. Vyas, "Some Aspects of Structural Change in Indian Agriculture", Indian Journal of Agricultural Economics, 34, 1, January-March 1979.

S. H. Wafa, "Land Development Strategies in Malaysia: An Empirical Study", Ph.D. Dissertation, Stanford University, July 1972.

D. Warriner. Land Reform in Principle and Practice. London: Oxford University Press, 1969.

C. R. Wharton, Jr., "The Infrastructure for Agricultural Growth". In Agricultural Development and Economic Growth, eds., H. M. Southworth and B. F. Johnston. Ithaca: Cornell University Press, 1967.

A. Wildavsky. Speaking Truth to Power: The Art and Craft of Policy Analysis. Boston and Toronto: Little, Brown and Company, 1979.

World Bank, Rural Enterprise and Nonfarm Employment. Washington, D.C.: World Bank, January 1978.

_____, World Development Report, 1980. New York: Oxford University Press, 1980.

Index